# Welfare of Cultured and Experimental Fishes

# Welfare of Cultured and Experimental Fishes

Special Issue Editors

**Pablo Arechavala-Lopez**
**Joao Luis Saraiva**

MDPI • Basel • Beijing • Wuhan • Barcelona • Belgrade

**MDPI**

*Special Issue Editors*
Pablo Arechavala-Lopez
Centro de Ciências do Mar
(CCMAR)
Portugal

Joao Luis Saraiva
Centro de Ciências do Mar
(CCMAR)
Portugal

*Editorial Office*
MDPI
St. Alban-Anlage 66
4052 Basel, Switzerland

This is a reprint of articles from the Special Issue published online in the open access journal *Fishes* (ISSN 2410-3888) from 2018 to 2019 (available at: https://www.mdpi.com/journal/fishes/special_issues/Fish_Welfare).

For citation purposes, cite each article independently as indicated on the article page online and as indicated below:

LastName, A.A.; LastName, B.B.; LastName, C.C. Article Title. *Journal Name* **Year**, *Article Number*, Page Range.

**ISBN 978-3-03921-710-6 (Pbk)**
**ISBN 978-3-03921-711-3 (PDF)**

Image courtesy of Pablo Arechavala-Lopez.

# Contents

# About the Special Issue Editors

**Pablo Arechavala-Lopez** (PhD). Pablo Arechavala-Lopez has a solid background in marine ecology, ichthyology, and coastal aquaculture. His research is mainly focused on the study of the behavior of cultured fish, as well as on the management of coastal areas, evaluating the influence of aquaculture activities on wild fish populations and coastal ecosystems. More recently, he has been applying his knowledge and experience to the field of fish welfare in aquaculture, analyzing the physiological and behavioral adaptations of fish to reared and enriched environments and also acting as a consultant on fish welfare for worldwide aquaculture companies.

**Joao L. Saraiva** (PhD). Joao L. Saraiva is a fish ethologist with a special interest in welfare. He is currently leading the Fish Ethology and Welfare Group at CCMAR and is the president of the FishEthoGroup Association. With a career built on the interplay between proximate and ultimate mechanisms underlying social behavior and communication in teleosts, Dr. Saraiva is now focusing on the application of fundamental science to improve the welfare of farmed fish. Along with his scientific activity, Dr. Saraiva is also part of the team of FishEthoBase, an open-access database on welfare of farmed fish species.

*fishes*

MDPI

*Editorial*

# Welfare of Fish—No Longer the Elephant in the Room

João L. Saraiva [1,*] and Pablo Arechavala-Lopez [1,2]

1   Fish Ethology and Welfare Group, CCMAR, 8000-536 Faro, Portugal
2   Fish Ecology Group, IMEDEA (CSIC/UIB), 07190 Esporles, Illes Balears, Spain
*   Correspondence: joao@fair-fish.net

Received: 27 June 2019; Accepted: 28 June 2019; Published: 3 July 2019

check for updates

The concept of fish welfare is fairly recent and was overlooked for many years, based on a popular misconception that fish were "stupid" creatures devoid of any kind of sentience or mental capability. However, a growing body of research on fish behaviour, cognition, learning and neuroscience made clear that this is evidently not the case—fish are indeed socially complex [1], have developed cognitive and learning abilities with the neural substrate to support them [2,3] and are sentient [4]. As the evidence for fish mental competences grew larger, the uncomfortable questions regarding the welfare state of fish and the ethical implications of fish farming became an elephant in the room that nobody would address. However, there are recent indications that things are changing, and the present collection of excellent contributions suggests that all the interested parties (scientists, farmers, retailers, NGOs and consumers) are now directly approaching the subject. Ladies and gentlemen, the elephant has left the building.

In fact, these papers are a fantastic example of the many perspectives that may be used when tackling fish welfare. In a pilot study regarding fisheries of catshark (*Scyliorhinus canicular*), Barragán-Méndez et al. [5] demonstrate that the standard practices of exposing the wild-caught animals to air are not only extremely harmful for the fish but also modify muscle texture properties and reduce the quality of the meat. This study demonstrates the urgency of improving the welfare of wild-caught fish and indicates the road ahead regarding the assessment of humane practices in fisheries. The paper by Strauch et al. [6] also highlights how a common practice of adding phosphate to integrated aquaponic systems as a fertilizer can have negative effects in African catfish (*Clarias gariepinus*), not only on their welfare but also on the meat quality. These two studies emphasize a correlation that should be clear for the fish industry: when the welfare of animals is improved, both the quality of the product and its value increase—a rare case when the interest of the industry and the ethical standards underlying its activity walk hand in hand.

The study by Moreira et al. [7] takes an ontogenic approach into amyloodiniosis, a well-known health problem in white seabream (*Diplodus sargus*) farmed in Southern Europe. Focusing on fish health is not new in welfare research. After all, health is one of the key components for conceptual framework of welfare, together with the mental and natural components. The novelty of this paper is to search for a non-veterinary approach to deal with a health issue, using one of Tinbergen's Four Questions that is so often overlooked: development. The results show that young fish are far more susceptible to infection by *Amyloodinium ocellatum* because they lack immune and physiological responses that only appear later in ontogeny. This study highlights the need to take into account the age of the individuals when designing prevention and treatment plans as well as rearing routines.

Zebrafish welfare was another surprisingly ignored issue until recent times. The number of cultured individuals arguably rivals any other commercially farmed species, yet even when the subject of welfare in aquaculture started to be addressed, zebrafish were apparently left behind. Say no more because two studies by Woodward et al. [8] and Deakin et al. [9] focus on two important topics that impact zebrafish welfare: the first shows that environmental enrichment in zebrafish housing promotes aggression and risk-taking behaviours in zebrafish [8], and the authors explain this with the social and territorial behaviour of the species, in which the enrichment structures provide resources to monopolise; the second suggests a novel method to analyse pain responses to standard experimental procedures in this species. Using fractal analysis of behaviour, the authors create (and validate) a pain intensity scale for zebrafish, and propose that variations in complexity of movement should be a good indicator of welfare in this species [9]. This paper also adds compelling evidence that fish are sentient and able to feel pain. Both studies dive into the biology of welfare in zebrafish, using basic behavioural variables and knowledge on the ethology of the species to highlight the importance of the natural (in the case of Woodward et al.) and mental (in the case of Deakin et al.) dimensions of welfare.

The importance of understanding the ethology of reared species is further explored in the review by Gonçalves-de-Freitas et al. [10], where the social behaviour of Nile tilapia (*Oreochromis niloticus*) is proposed as a key component in the welfare of this fish. In this study, the authors thoroughly review the social ethology of tilapia, elegantly addressing both proximate and ultimate mechanisms to provide operational insights that may improve its welfare. The social environment is demonstrated to have impacts on stress levels, growth and aggression, and the authors offer solutions to mitigate the effects of rearing conditions: lighting, environment colour and enrichment structures are pinpointed as simple ways to reduce the detrimental effects of human-induced social disturbance [10].

The review by Fife-Cook and Franks [11] picks up where Gonçalves-de-Freitas et al. left off. In their paper, the authors propose a framework for positive welfare in fish (i.e., mental and physical states that exceed what is necessary for immediate survival), which would replace the traditional paradigms that focus on mitigating the negative impacts of rearing. The positive welfare approach requires a deep understanding of the species' behaviour and biology and would demand taxa-specific standards. However, the knowledge already available from fish and other taxa allows both the identification of positive welfare states in fish and the suggestion of active measures: species-specific housing (including ambient colour and appropriate social environment, as already suggested in [10]) and the promotion of cognitive engagement (visual stimulation, novel objects, play, etc.). The authors conclude that the positive class of experiences are currently being neglected in fish and urge for more research in this area.

Finally, Saraiva et al. [12] propose a framework to assess the welfare of farmed fishes at a species level: the FishEthoBase. This open-access database on fish ethology and welfare aims to provide a tool to evaluate the welfare state of (ultimately) all farmed species worldwide. In that sense, the authors have built the portal www.fishethobase.net where an impressive amount of data concerning the biology of farmed species and the impacts of aquaculture on their welfare is scrutinised, organised and summarised. Using the data on 41 species already available, the authors analyse their welfare state and conclude that (i) the general welfare state of farmed fishes is poor, (ii) there is some potential for improvement, and (iii) this potential is related to research on species' needs, but (iv) there are many remaining knowledge gaps, and (v) current fish farming technologies do not seem to fully address welfare issues.

As editors of this Special Issue, we could not be more thankful and proud of its contents, not only because of the quality of the contributions but also because of their broad approach, their interconnection and the multiple doors that have opened, letting us see future lines of knowledge. Fish welfare seems to have gained a considerable momentum and, although there is yet much work ahead, we can optimistically say that the wind is blowing in a favourable direction.

**Conflicts of Interest:** The authors declare no conflict of interest

## References

1. Oliveira, R.F. Social plasticity in fish: Integrating mechanisms and function. *J. Fish Biol.* **2012**, *81*, 2127–2150. [CrossRef] [PubMed]
2. Bshary, R.; Gingins, S.; Vail, A.L. Social cognition in fishes. *Trends Cogn. Sci.* **2014**, *18*, 465–471. [CrossRef] [PubMed]

3. Oliveira, R.F. Mind the fish: Zebrafish as a model in cognitive social neuroscience. *Front. Neural Circuits* **2013**, *7*. [CrossRef] [PubMed]

4. Brown, C. Fish intelligence, sentience and ethics. *Anim. Cogn.* **2015**, *18*, 1–17. [CrossRef] [PubMed]

5. Barragán-Méndez, C.; Sánchez-García, F.; Sobrino, I.; Mancera, J.M.; Ruiz-Jarabo, I. Air Exposure in Catshark (Scyliorhinus canicula) Modify Muscle Texture Properties: A Pilot Study. *Fishes* **2018**, *3*, 34. [CrossRef]

6. Strauch, S.M.; Bahr, J.; Baßmann, B.; Bischoff, A.A.; Oster, M.; Wasenitz, B.; Palm, H.W. Effects of Ortho-Phosphate on Growth Performance, Welfare and Product Quality of Juvenile African Catfish (Clarias gariepinus). *Fishes* **2019**, *4*, 3. [CrossRef]

7. Moreira, M.; Cordeiro-Silva, A.; Barata, M.; Pousão-Ferreira, P.; Soares, F. Influence of Age on Stress Responses of White Seabream to Amyloodiniosis. *Fishes* **2019**, *4*, 26. [CrossRef]

8. Woodward, M.A.; Winder, L.A.; Watt, P.J. Enrichment Increases Aggression in Zebrafish. *Fishes* **2019**, *4*, 22. [CrossRef]

9. Deakin, A.G.; Spencer, J.W.; Cossins, A.R.; Young, I.S.; Sneddon, L.U. Welfare Challenges Influence the Complexity of Movement: Fractal Analysis of Behaviour in Zebrafish. *Fishes* **2019**, *4*, 8. [CrossRef]

10. Gonçalves-de-Freitas, E.; Bolognesi, M.C.; dos Santos Gauy, A.C.; Brandão, M.L.; Giaquinto, P.C.; Fernandes-Castilho, M. Social Behavior and Welfare in Nile Tilapia. *Fishes* **2019**, *4*, 23. [CrossRef]

11. Fife-Cook, I.; Franks, B. Positive Welfare for Fishes: Rationale and Areas for Future Study. *Fishes* **2019**, *4*, 31. [CrossRef]

12. Saraiva, J.L.; Arechavala-Lopez, P.; Castanheira, M.F.; Volstorf, J.; Heinzpeter Studer, B. A Global Assessment of Welfare in Farmed Fishes: The FishEthoBase. *Fishes* **2019**, *4*, 30. [CrossRef]

*fishes*

MDPI

Article

# Air Exposure in Catshark (*Scyliorhinus canicula*) Modify Muscle Texture Properties: A Pilot Study

Cristina Barragán-Méndez [1], Fini Sánchez-García [2], Ignacio Sobrino [3], Juan Miguel Mancera [1] and Ignacio Ruiz-Jarabo [1,*]

[1] Departament of Biology, Faculty of Marine and Environmental Sciences, Instituto Universitario de Investigación Marina (INMAR), Universidad de Cádiz, Campus de Excelencia Internacional del Mar (CEI-MAR), Av. República Saharaui s/n, E-11510 Puerto Real, Cádiz, Spain; cristina.barragan@uca.es (C.B.-M.); juanmiguel.mancera@uca.es (J.M.M.)

[2] Departament of Chemical Engineering and Food Technology, Faculty of Sciences, Universidad de Cádiz, Campus de Excelencia Internacional del Mar (CEI-MAR), Av. República Saharaui s/n, E-11510 Puerto Real, Cádiz, Spain; fini.sanchez@uca.es

[3] Instituto Español de Oceanografía (IEO), Centro Oceanográfico de Cádiz, Puerto Pesquero, Muelle de Levante, s/n, P.O. Box 2609, E-11006 Cádiz, Spain; ignacio.sobrino@ieo.es

* Correspondence: ignacio.ruizjarabo@uca.es; Tel.: +34-615514755

Received: 4 July 2018; Accepted: 31 August 2018; Published: 4 September 2018

check for updates

**Abstract:** Sharks are captured by tons for human consumption. Improving the quality of their meat will produce fillets that may have a higher economic value in the market, and thus be beneficial for the management of this fishery. In other animal species destined for human consumption, a negative relationship between pre-slaughtering stress and meat quality has been demonstrated. By studying the commercial small-spotted catshark (*Scyliorhinus canicula*), this work aimed at linking pre-slaughter handling of captured sharks and muscle fillets quality. An experimental group of adult and subadult living catsharks captured by hand and exposed to air (for 18 min, which is the minimum time this species is exposed to air in the fishing deck during fisheries procedures), and an undisturbed group, were evaluated. After air exposure, catsharks returned to water for recovery. Muscle lactate and water content were quantified after acute exposure (for 18 min), 5 h and 24 h. This challenge elicited stress responses in the muscle such as increased lactate levels and immediate dehydration, followed by recovery of lactate levels and overhydration. Muscle consistency, a relevant variable describing quality of seafood according to its ability to be swallowed by the consumer, paralleled muscle water content changes. The results indicated for the first time that handling alive sharks exposed to air results in muscle fillets with different texture properties. Whether these changes in muscle texture induce higher quality fillets has yet to be proven. Our recommendation is to minimize time of air exposure experienced by sharks when captured, including fast slaughtering instead of leaving them to die by asphyxia, as current on-board procedures.

**Keywords:** fisheries management; muscle texture; *Scyliorhinus canicula*; sharks; stress

## 1. Introduction

World consumption of elasmobranchs has increased during the last years, reaching 121,641 tons in 2011, which represents an increase of 42% by volume compared with 2000 [1]. According to the Food and Agriculture Organization of the United Nations, Spain is the world's third-largest producer of shark meat, and capture statistics show a strong upward trend in recent years [1].

The quality of the flesh in different species depends on the handling method of the fish before being slaughtered [2]. Improving the quality of seafood is a topical issue in the industry. Thus, slaughter handling stress is important for the aquaculture industry because of its effects on fillet quality [3]. The degree of muscle texture degradation is related to the length of the pre-slaughter stress period in teleosts [4]. Moreover, stressful stunning methods also induce a higher deterioration rate in texture properties as seen in silver carp [4]. Fish flesh quality is a very complex concept that includes physico-chemical, biochemical, visual (color of the fillets) and microbiological attributes [5]. The use of texture analysis is widely extended in the food industry for the assessment of teleost fish fillet quality [3,6], as it provides more objective measurements [7]. Texture profile analysis test is based on the imitation of mastication or chewing process, and is employed for the evaluation of seafood, mostly teleost fish [8]. In this sense, texture consistency evidenced the water holding capacity of the fillet and reflects the difficulty of swallowing a piece of it [9]. However, despite the large volume of sharks captured for human consumption, little is known regarding their meat quality and texture properties. Stress in fish elicits physiological responses to cope with the environmental disturbance (presence of a predator, capture by a hook or a net, air exposure, etc.). These responses are broadly grouped as primary, secondary and tertiary [10]. Primary responses include the release of hormones to the blood, being cortisol for teleost fish and (presumably) $1\alpha$-hydroxycorticosterone in elasmobranchs [11–13]. Secondary responses are defined as the immediate actions and effects of these hormones [14], which basically make available to the cells an increased supply of oxygen and energy metabolites. These responses involve modifications in heart and breathing rates [15,16], changes in the dilation status of blood vessels [17], and rapid mobilization of glucose to meet energetic demands [18,19]. The onset of anaerobic glycolysis accumulates lactate in white muscle of teleosts [20]. As a side effect, dehydration processes occur in seawater elasmobranchs, enhancing plasma osmolality and sodium content after capture [21–23]. All these responses have consequences on the muscle texture properties of teleost fish [24,25], although no information is available regarding pre-handling stress and changes in muscle texture in elasmobranchs.

The small-spotted catshark (*Scyliorhinus canicula*) is a small demersal species inhabiting Atlantic waters from Norway to Senegal and the Mediterranean Sea, mostly captured by bottom trawl fisheries [26]. This and other catshark species are particularly appreciated in Europe [1]. In Spain, trawling time ranges from 1 h to 4 h [27], but this species may be exposed to air for 3 h before dying (personal observation). Thus, the objective of the present study was to use already established knowledge in sharks and teleost regarding stress reactions, and investigate if the relationship known from teleost, with flesh quality (based on muscle texture properties), also holds true in sharks. This information will be useful to examine the effects of exposure to air on the flesh quality (muscle texture) in the absence of the effects of prior fishing-related capture in the small-spotted catshark.

## 2. Results and Discussion

No mortality was observed in any group throughout the experimental period. All animals employed in this study (November 2016) evidenced signs of good health along the experimental time (from capture by a bottom-trawler vessel, to being euthanized at the end of the experiment), with normal behavior inside the tanks, and without external or internal evidence of disease. *P*-values resulting from two way-ANOVA of the different parameters assessed are presented in Table 1.

**Table 1.** *P*-values from two way-ANOVA from General Lineal Model (GLM) of parameters measured in muscle of *Scyliorhinus canicula* exposed to undisturbed control conditions or after an acute capture by hand processes followed by air exposure, and sampled at times 0, 5 and 24 h after the challenge.

| Parameter | Group | Time | Group × Time |
|---|---|---|---|
| Lactate | <0.0005 | <0.000001 | <0.0001 |
| Water content | <0.05 | <0.000001 | <0.00001 |
| Consistency | 0.7703 | <0.0001 | <0.0001 |
| Cohesiveness | 0.5715 | 0.6419 | 0.7598 |
| Crispiness | 0.7509 | 0.6896 | 0.6139 |
| Firmness | 0.6939 | 0.1156 | 0.8871 |
| Work of penetration | 0.6317 | 0.1452 | <0.005 |
| Springiness | 0.1226 | <0.0005 | 0.2542 |

## *2.1. Muscle Lactate*

*Scyliorhinus canicula* when captured under normal bottom trawl fisheries conditions experienced periods of air exposure around 18 min or more [26,28]. Our study confirms that capture by hand followed by air exposure acutely stressed the animals as muscle lactate levels increased significantly (Figure 1), due to increased anaerobic energy metabolism [24,25]. In addition, our results are similar to those described for gummy shark (*Mustelus antarcticus*) captured by gill-net [29], with highest muscle lactate levels just after the stress situation induced by capture, decreasing to control values after 3–5 h.

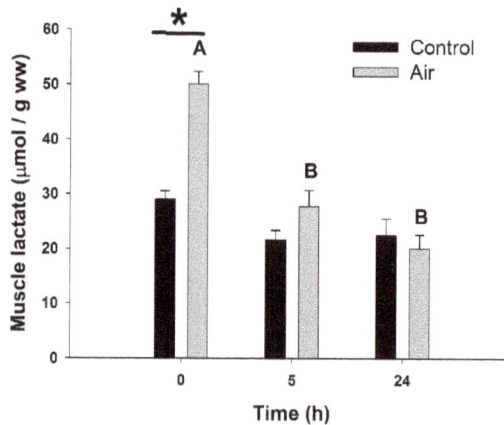

**Figure 1.** Muscle lactate levels (in μmol lactate per g of wet weight) in small-spotted catshark (*S. canicula*) controls (black bars) and after capture by hand and acute air exposure (gray bars). Data are expressed as mean ± standard error of the mean (SEM). No differences were described in the control group over time. Different letters indicate significantly different groups for the same treatment, while asterisks (*) indicate differences between control and air-exposure groups for the same time ($p < 0.05$, two-way ANOVA followed by a Tukey's post hoc test, $n = 7$–8).

## *2.2. Muscle Water Content*

The percent of water in the muscle in this study ranged between 71.8% and 78.8% water (Figure 2A), in accordance with the water holding capacity (WHC) described for other shark species captured by longline (60–80% WHC) [30]. Figure 2A shows muscle dehydration after 18 min of capture by hand and air exposure (0 h). Muscle dehydration could be associated with a secondary stress response in fish maintained in seawater and it could be related to the described increase in plasma osmolality in *Solea senegalensis* after an acute air-exposure situation [31], or to the highest plasma potassium concentration in Port Jackson sharks (*Heterodontus portusjacksoni*) and gummy sharks after gill-net capture [32]. This dehydration was described as a secondary stress response due to enhanced gill permeability, with an efflux of water to the hyperosmotic seawater environment. However, the apparent dehydration process in air-exposed catsharks 18 min after the challenge, sampled without being returned into the water, require further explanation. Thus, it was described that endurance exercise (similar to that performed by catsharks struggling for 18 min while exposed to air) induced a $\geq$2% body mass loss due to extracellular fluid volume contraction and, thus, muscle dehydration [33]. This body mass loss may be related to the reduction in muscle water content described in our study, with 74.5% and 73.6% muscle water content in the control and the stressed groups, respectively, at time 0 h after the stressful challenge. In this study, muscle water content enhanced in the experimental group after 5 h, suggesting the existence of a rebound effect due to the activation of the osmoregulatory system to recover the homeostatic balance after the stressful situation induced by air exposure. Several authors have studied the consequences of an acute stress on the osmoregulatory mechanisms in elasmobranchs and teleost fish showing a similar situation but different recovery times [31,34,35]. Our results showed that the small-spotted catshark was able to recover the muscle water content of the control-unstressed group within the first 24 h post-stress.

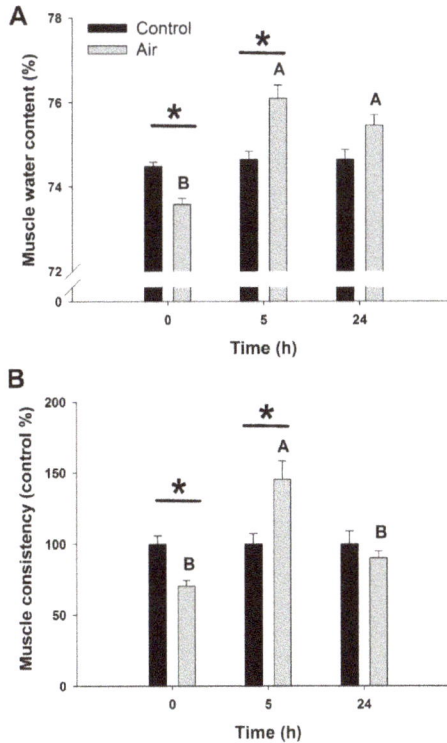

**Figure 2.** Muscle (**A**) water content (percent of total wet weight) and (**B**) consistency (as percent of variation compared to the control group at each sampling time) in small-spotted catshark (*S. canicula*) controls (black bars) and after capture by hand and acute air exposure (gray bars). Data are expressed as mean ± SEM. No differences were described in the control group over time. Different letters indicate significantly different groups for the same treatment, while asterisks (*) indicate differences between control and air-exposure groups for the same time ($p < 0.05$, two-way ANOVA followed by a Tukey's post hoc test, $n = 7$–8).

*2.3. Muscle Texture*

The most relevant result of the present study is that muscle consistency (Figure 2B), amongst all the texture-related parameters analyzed (Table 2), seems to be related to muscle water content. Fish muscle basically consists of connective tissues (including the extracellular aqueous matrix) and the intracellular contractile proteins. To determine consistency, the fillet has to be compressed producing a loss of intrafibrillar integrity, especially affecting actomyosin, which may influence the consistency/firmness values [36]. Muscle consistency roughly describes the capacity of the tissue to retain water after being compressed, and is useful to determine the swallowing properties of it [9]. It was described in salmon that muscle firmness was related not only to total collagen content but also to collagen stability [37]. As no differences were found in muscle firmness but in consistency in the present study, we can assume that both collagen content and stability were not affected by capture by hand and air exposure in *S. canicula*, being water content and the water holding capacity the most affected parameters during this study. Thus, major changes occurred in the extracellular matrix, while the structural integrity of the intracellular proteins is maintained after air exposure, as supported by the lack of differences in the analyzed texture parameters of this study (Table 2). The increased muscle consistency after 5 h recovery in this study may be related to an osmoregulatory rebound effect, but we were unable to find similar results in the literature. As far as we know, the shrinkage of shark fillets (due to denser myofibrillar and connective tissues) results in badly appreciated tough fillets [38]. Based on our results, if we considered that low muscle consistency is linked to dehydration and therefore tougher fillets, further studies should be conducted to elucidate if higher muscle consistency has better organoleptic properties. It should also be mentioned that freezing–thawing will itself affect muscle texture, but sometimes during commercial fisheries captured sharks are frozen aboard the vessel, and thus a possible bias of the results is assumed.

**Table 2.** Muscle texture profile analysis of small-spotted catshark (*S. canicula*) controls (Control) and after capture by hand and acute air exposure (Air). Data are expressed as mean $\pm$ SEM. The asterisk (*) indicate differences between control and air-exposure groups for the same time; no more differences have been described between groups or over time ($p < 0.05$, two-way ANOVA followed by a Tukey's post hoc test, $n = 7$–8).

| Parameter | Group | 0 h | 5 h | 24 h |
|---|---|---|---|---|
| Cohesiveness (N) | Control | $-11.9 \pm 2.6$ | $-11.9 \pm 4.0$ | $-10.5 \pm 1.1$ |
| | Air | $-8.8 \pm 2.0$ | $-12.6 \pm 1.6$ | $-9.5 \pm 2.7$ |
| Crispiness (no units) | Control | $199 \pm 3$ | $195 \pm 5$ | $197 \pm 4$ |
| | Air | $197 \pm 2$ | $198 \pm 5$ | $193 \pm 4$ |
| Firmness (N) | Control | $244 \pm 2$ | $246 \pm 1$ | $246 \pm 1$ |
| | Air | $242 \pm 1$ | $247 \pm 2$ | $245 \pm 2$ |
| Work of penetration (N mm) | Control | $257 \pm 7$ | $228 \pm 12$ | $280 \pm 15$ |
| | Air | $234 \pm 14$ | $284 \pm 14$ * | $260 \pm 10$ |
| Springiness (mm) | Control | $-38.4 \pm 1.9$ | $-36.2 \pm 2.9$ | $-28.1 \pm 2.5$ |
| | Air | $-40.4 \pm 3.5$ | $-29.0 \pm 4.2$ | $-20.9 \pm 3.9$ |

It was previously described in another shark species, *Mustelus lunulatus*, that postmortem changes such as loss of fluid and reduction of water holding capacity occurred in the muscle, but no changes in its texture were described [39]. The WHC has been reported as a good indicator for fish quality evaluation, being its reduction associated to tougher fillets and a loss of myofibrillar proteins [39]. Firmness of fish muscle is influenced by many factors [6,38], and the use of a texture profile analyzer seems to be useful when studying sensory conditions of raw fish fillets [40]. In this sense, further studies related to fillet firmness of elasmobranch fish including variables such as size, sex and fishing gear, amongst other variables, will be relevant to better describe their effects on the quality of the meat. Thus, Atlantic salmon fillets were firmer, paler, had a higher muscle pH and lower liquid loss in fish that were not stressed during pre-slaughter handling [41]. Moreover, color variations in the meat serves as a quality standard for seafood products. Using this parameter, it was described that exhausted salmon [42] and stressed Bluefin tuna [2] showed changes in fillet color. Thus, it should be interesting to further analyze fillet color in future studies involving sharks employed for human consumption.

As this was designed as a pilot study, where we aimed at testing the effects of acute stress on the textural properties of the muscle using as few animals as possible, some flaws were evidenced. Future studies should be designed carefully and be focused in the individual stressful processes that sharks may experience when captured (such as muscle exhaustion after swimming, confinement, hypoxic conditions, blows, changes in the environmental conditions and, finally, air exposure alone), presumably including more animals per experimental group.

## 3. Material and Methods

### 3.1. Animal

*Scyliorhinus canicula* adults and subadults of both sexes ($n$ = 26 males and $n$ = 19 females, ranging from 169 to 512 g body wet weight and 41.5 to 57.0 cm total length) were obtained by bottom trawling in November 2016 in the Gulf of Cadiz (Spain). Healthy animals (with no external wounds, normal skin color and swim behavior in the onboard tanks employed to recover captured animals after trawling, active responses to external stimuli such as body movement when other catsharks touched them, and eyes clear and reactive) were transported immediately (transport process lasted around 8–12 h) to the fish husbandry facility of the Faculty of Marine and Environmental Sciences (Puerto Real, Cadiz, Spain) and acclimated to seawater (salinity 38 psu) until the beginning of the experiment. Fish were randomly divided into 6 bare-bottom, rectangular soft-edges tanks of 400 L (surface area of 0.72 m$^2$, 56 cm depth, and covered by a fine-mesh tissue to shade the aquarium) in a flow-through system under natural photoperiod (November; latitude 36°31′34″ N) and temperature (ambient temperature of approximately 19 °C, while temperature of the seawater when captured was 18.81–21.19 °C) and acclimated for 17 days. The number of males and females in each tank, though randomly distributed, included 4–5 males and 3–4 females per sampling group. Fish density per unit of surface was in accordance with previous studies conducted in *Squalus acanthias* and *Squalus suckleyi* [35,43]. Physical-chemical parameters were kept stable along the experimental time by means of the constant flow-through of well-seawater (constant 38 psu salinity, temperature 19.0 °C, >5 mg Dissolved Oxygen or DO L$^{-1}$, $0.1 \pm 0.2$ mg Total Ammonia Nitrogen or TAN L$^{-1}$, $0 \pm 4$ µg NH$_3$ L$^{-1}$, <0.1 mg nitrites L$^{-1}$ and <4.1 mg nitrates L$^{-1}$). Salinity, temperature and dissolved oxygen levels were daily measured inside each tank. Temperature oscillated inside the tanks between 19.0 and 19.3 °C along the day. Oxygen levels were daily measured with an oximeter (Handy Polaris, OxyGuard, DK) and maintained above 5 mg L$^{-1}$ (around 90% saturation) by an air diffusing stone inside each tank. Nitrogen compounds (TAN, NH$_3$, nitrites and nitrates) were analyzed once a week in the incoming well-water (which is also daily analyzed by the people in charge of the Fish Husbandry facility, without further differences along time) with commercial kits (Merck, Germany). Health of the animals was visually assessed, taking special interest in controlling that the body and eyes color, breathing rates and behavior of all sharks were normal. Fish were fed once a day at 20.00 UTC with unfrozen shrimps, prawns, sardines and anchovies to satiety. Red lights were employed for feeding in order to not disturb the animals during this process. As a curiosity, the animals remained inactive at the bottom of the tank most of the time but, after a few days, at the time of feeding began to actively look for the food offered, reaching to catch it directly from the hand of the caregiver with tranquility. This fact can be taken, with caution, as a sign that the animals were not specially stressed. Fecal matter was carefully removed each morning with a siphon under ambient darkened conditions. Animals were fasted 36 h before sampling to avoid osmoregulatory imbalances related to feeding, as has been described before in other catshark species [44]. All experimental procedures complied with the Guidelines of the European Union (2010/63/UE) and the Spanish legislation (RD 1201/2005 and law 32/2007) for the use of laboratory animals, and were approved by the local Committee of Ethics and Animal Experimentation.

*3.2. Experimental Design and Sampling*

Capture and air exposure are dramatic stressful events that catshark experience during fisheries procedures. Therefore, three tanks were selected as undisturbed controls, and catshark from the other three tanks were captured and exposed to the air. Animals were captured by hand and placed in similar dry tanks for 18 min, which is the minimum time this species experienced outside water during commercial bottom-trawl fishing conditions [26,28]. We considered unnecessary to fully expose the animals to the exact conditions after commercial fishing. Samples were collected at times 0 and 24 h after air exposure, consisting of 2–3 animals from each tank (in triplicate, $n = 7$–8 per group). The group sampled 0 h after the air exposure was not introduced back into the water tanks after the challenge. Moreover, an additional sampling point after 5 h was conducted to evaluate recovery responses according to previous studies in *S. canicula* and *Scyliorhinus stellaris* [45,46]. First sampling for both experimental groups was performed at 08:30–09:30 UTC. Catsharks were captured by hand and immediately anesthetized in 0.1% $v/v$ 2-phenoxiethanol (P-1126, Sigma-Aldrich, St. Louis, MI, USA). Unless anesthesia has been shown to affect stress-related blood variables in sharks [47] we decided to anesthetize the animals and accept possible bias in the muscle parameters as part of this pilot study. Weight and length of the animals was measured. Euthanize was done by severing the head with a sharp knife. A 5-cm portion of the trunk, covering the area just after the first dorsal fin, was collected and immediately frozen at $-20\ ^{\circ}$C. The process of freezing the trunk was previously described in similar studies [30,38] and resembles commercial activities for marketed *S. canicula* in the South of Spain. All procedures lasted less than four minutes per tank, aiming at minimizing secondary-stress responses due to fish manipulation [10].

*3.3. Muscle Analysis*

The samples were allowed to unfreeze overnight at 4 °C, the skin was removed, and the muscle fillets from both sides of the spine were separated by using a sharp knife. Previous to overnight unfreezing, muscle lactate was analyzed in 0.2 g subsamples (still at $-20\ ^{\circ}$C) from these fillets as previously described [48]. In brief, frozen muscle was finely minced on an ice-cooled Petri dish, homogenized by ultrasonic disruption in 7.5 volumes ice-cold 0.6 N perchloric acid, neutralized using 1 M potassium bicarbonate, centrifuged (3 min at 10,000 g, Eppendorf 5415R), and the supernatant used to assay muscle lactate. Lactate was determined spectrophotometrically with a commercial kit adapted for 96-well microplates (ref. 1001330, Spinreact, St. Esteve de Bas, Girona, Spain).

Square pieces (3 cm × 3 cm) of each muscle fillet were cut and maintained on an ice-cold Petri dish for texture analysis. Instrumental texture analyses were performed on the 2 fillets per fish (in duplicate) using a texture analyzer TA1 (Lloyd-instruments, AMETEK GmbH, Meerbusch, Germany) equipped with an 80 N compression load cell controlled with the program Nexygen Plus 3.0 for Windows. A flat-ended cylindrical probe (12.5 mm diameter; type P/0.5) was used to analyze the consistency, cohesiveness, crispiness, firmness, work of penetration and springiness of the muscle fillets. Analyses were performed by pressing the cylinder into the muscle fillet at a constant crosshead speed of 4 mm s$^{-1}$. All samples were compressed twice to 50% of their original height. The parameters analyzed (and their Units) were cohesiveness (in N), consistency (in N mm), crispiness (no Units), firmness (in N), work of penetration (in N mm) and springiness (in mm). Cohesiveness describes how well a food retains its form between the first and second chew (is expressed as the area of work during the second compression divided by the area of work during the first compression). Consistency relates to the firmness of a semisolid, and is calculated as the breaking force in the first compression (N) multiplied by the deformation between first and second compression (mm). Crispiness or fracturability is the tendency of a material to fracture, crumble, crack, shatter or fail upon the application of a relatively small amount of force or impact. Firmness is related to the resistance to deformation (given by the peak load of the initial compression). Work of penetration is described as the distance (in mm) necessary to give the maximum strength (in N) after the initial compression. Springiness is how well a product physically springs back after it has been deformed during the first compression and has been allowed to wait for the target wait time between strokes (expressed as the distance of the detected height during the second compression divided by the original compression distance). Muscle water content was analyzed by drying preweighted muscle at 65 °C until constant weight (circa 72 h), as previously described [49]. The percentage of water was calculated as the difference in weight between the fresh and the dry muscle divided by the fresh weight.

*3.4. Statistics*

Normality and homogeneity of variances were analyzed using the Shapiro–Wilk's test and the Levene's test, respectively. Differences between groups were tested using two-way ANOVA with group (control and capture by hand/air exposure) and time (0, 5 and 24 h) as the factors. When ANOVA yielded significant differences, Tukey's post-hoc test was used to identify significantly different groups. Statistical significance was set at $p < 0.05$. All results are given as mean $\pm$ SEM. All tests were performed using Statistica 7 software for Windows.

# 4. Conclusions

We have described, for the first time, that elasmobranch flesh texture conditions are affected by pre-slaughtering handling and/or air exposure of the animals. However, future studies are necessary to better characterize fillet quality in sharks, especially after fisheries procedures.

**Author Contributions:** I.R.-J., I.S. and C.B.-M. designed the experiment. I.R.-J., C.B.-M. and F.S.-G. conducted the animal maintenance and analysis. I.R.-J., C.B.-M., F.S.-G., J.M.M. and I.S. wrote the manuscript.

**Funding:** This work was partially supported by funds from the PADI Foundation (App # 28467) to I.R.-J.; projects AGL2013-48835-C2-R and AGL2016-76069-C2-1-R (Ministerio de Economía y Competitividad); and project SUREDEPAR (Programa Pleamar, Ministerio de Agricultura y Pesca, Alimentación y Medio Ambiente) to J.M.M.

**Acknowledgments:** The authors are indebted to Emilio Garcia-Robledo for the correction of grammatical errors and English improvement. We would like to thank Sara Mohamed, Arturo Jiménez, Borja Carcas, Miriam Fernández, and the crew of the Miguel Oliver research vessel for their help during the sampling of the animals.

**Conflicts of Interest:** The authors declare no competing or financial interest. The funding sponsors had no role in the design of the study; in the collection, analyses, or interpretation of data; in the writing of the manuscript, and in the decision to publish the results.

# References

1. Dent, F.; Clarke, S. *State of the Global Market for Shark Products*; Food and Agriculture Organization of the United Nations: Rome, Italy, 2015; Volume 590, p. 196.
2. Addis, P.; Secci, M.; Locci, I.; Cau, A.C. Harvesting, handling practices and processing of bluefin tuna captured in the trap fishery: Possible effects on the flesh quality. *Collect. Vol. Sci. Pap. ICCAT* **2012**, *67*, 390–398.
3. Bahuaud, D.; Morkore, T.; Langsrud, O.; Sinnes, K.; Veiseth, E.; Ofstad, R.; Thomassen, M.S. Effects of −1.5 °C super-chilling on quality of Atlantic salmon (*Salmo salar*) pre-rigor fillets: Cathepsin activity, muscle histology, texture and liquid leakage. *Food Chem.* **2008**, *111*, 329–339. [CrossRef] [PubMed]
4. Zhang, L.; Li, Q.; Lyu, J.; Kong, C.; Song, S.; Luo, Y. The impact of stunning methods on stress conditions and quality of silver carp (*Hypophthalmichthys molitrix*) fillets stored at 4 °C during 72 h postmortem. *Food Chem.* **2017**, *216*, 130–137. [CrossRef] [PubMed]
5. Barat, J.M.; Gil, L.; Garcia-Breijo, E.; Aristoy, M.C.; Toldra, F.; Martinez-Manez, R.; Soto, J. Freshness monitoring of sea bream (*Sparus aurata*) with a potentiometric sensor. *Food Chem.* **2008**, *108*, 681–688. [CrossRef] [PubMed]

6.  Dong, X.P.; Wu, Q.; Li, D.Y.; Pan, J.F.; Zheng, J.J.; Fu, X.X.; Qi, L.B.; Chen, G.B. Physicochemical, micro-structural, and textural properties of different parts from farmed common carp (*Cyprinus carpio*). *Int. J. Food Prop.* **2017**, *20*, 946–955. [CrossRef]
7.  Chen, L.; Opara, U.L. Texture measurement approaches in fresh and processed foods—A review. *Food Res. Int.* **2013**, *51*, 823–835. [CrossRef]
8.  Cheng, J.H.; Sun, D.W.; Zeng, X.A.; Liu, D. Recent advances in methods and techniques for freshness quality determination and evaluation of fish and fish fillets: A review. *Crit. Rev. Food Sci. Nutr.* **2015**, *55*, 1012–1225. [CrossRef] [PubMed]
9.  Taniguchi, H.; Tsukada, T.; Ootaki, S.; Yamada, Y.; Inoue, M. Correspondence between food consistency and suprahyoid muscle activity, tongue pressure, and bolus transit times during the oropharyngeal phase of swallowing. *J. Appl. Physiol. (1985)* **2008**, *105*, 791–799. [CrossRef] [PubMed]
10. Barton, B.A. Stress in fishes: A diversity of responses with particular reference to changes in circulating corticosteroids. *Integr. Comp. Biol.* **2002**, *42*, 517–525. [CrossRef] [PubMed]
11. Hazon, N.; Henderson, I.W. Secretory dynamics of 1α-hydroxycorticosterone in the elasmobranch fish, *Scyliorhinus canicula*. *J. Endocrinol.* **1984**, *103*, 205–211. [CrossRef] [PubMed]
12. Armour, K.J.; O'Toole, L.B.; Hazon, N. The effect of dietary protein restriction on the secretory dynamics of 1α-hydroxycorticosterone and urea in the dogfish, *Scyliorhinus canicula*: A possible role for 1α-hydroxycorticosterone in sodium retention. *J. Endocrinol.* **1993**, *138*, 275–282. [CrossRef] [PubMed]
13. Iwama, G.K. Stress in fish. *Ann. N. Y. Acad. Sci.* **1998**, *851*, 304–310. [CrossRef]
14. Mommsen, T.P.; Vijayan, M.M.; Moon, T.W. Cortisol in teleosts: Dynamics, mechanisms of action, and metabolic regulation. *Rev. Fish Biol. Fish.* **1999**, *9*, 211–268. [CrossRef]
15. Wendelaar Bonga, S.E. The stress response in fish. *Physiol. Rev.* **1997**, *77*, 591–625. [CrossRef] [PubMed]
16. Shadwick, R.E.; Farrell, A.P.; Brauner, C. *Physiology of Elasmobranch Fishes: Internal Processes*; Academic Press: London, UK, 2015; Volume 34B, p. 580.
17. Theoharides, T.C.; Singh, L.K.; Boucher, W.; Pang, X.; Letourneau, R.; Webster, E.; Chrousos, G. Corticotropin-releasing hormone induces skin mast cell degranulation and increased vascular permeability, a possible explanation for its proinflammatory effects. *Endocrinology* **1998**, *139*, 403–413. [CrossRef] [PubMed]
18. Martins, C.L.; Walker, T.I.; Reina, R.D. Stress-related physiological changes and post-release survival of elephant fish (*Callorhinchus milii*) after longlining, gillnetting, angling and handling in a controlled setting. *Fish. Res.* **2018**, *204*, 116–124. [CrossRef]
19. Skrzynska, A.K.; Maiorano, E.; Bastaroli, M.; Naderi, F.; Miguez, J.M.; Martinez-Rodriguez, G.; Mancera, J.M.; Martos-Sitcha, J.A. Impact of air exposure on vasotocinergic and isotocinergic systems in gilthead sea bream (*Sparus aurata*): New insights on fish stress response. *Front. Physiol.* **2018**, *9*, 15. [CrossRef] [PubMed]
20. Skomal, G.B.; Mandelman, J.W. The physiological response to anthropogenic stressors in marine elasmobranch fishes: A review with a focus on the secondary response. *Comp. Biochem. Physiol. A* **2012**, *162*, 146–155. [CrossRef] [PubMed]
21. Hoffmayer, E.R.; Parsons, G.R. The physiological response to capture and handling stress in the Atlantic sharpnose shark, *Rhizoprionodon terraenovae*. *Fish Physiol. Biochem.* **2001**, *25*, 277–285. [CrossRef]
22. Mandelman, J.W.; Farrington, M.A. The physiological status and mortality associated with otter trawl capture, transport, and captivity of an exploited elasmobranch, *Squalus acanthias*. *ICES J. Mar. Sci.* **2007**, *64*, 122–130. [CrossRef]
23. Brill, R.; Bushnell, P.; Schroff, S.; Seifert, R.; Galvin, M. Effects of anaerobic exercise accompanying catch-and-release fishing on blood-oxygen affinity of the sandbar shark (*Carcharhinus plumbeus*, Nardo). *J. Exp. Mar. Biol. Ecol.* **2008**, *354*, 132–143. [CrossRef]
24. Bahuaud, D.; Morkore, T.; Ostbye, T.K.; Veiseth-Kent, E.; Thomassen, M.S.; Ofstad, R. Muscle structure responses and lysosomal cathepsins b and l in farmed Atlantic salmon (*Salmo salar* L.) pre- and post-rigor fillets exposed to short and long-term crowding stress. *Food Chem.* **2010**, *118*, 602–615. [CrossRef]
25. Poli, B.M.; Parisi, G.; Scappini, F.; Zampacavallo, G. Fish welfare and quality as affected by pre-slaughter and slaughter management. *Aquac. Int.* **2005**, *13*, 29–49. [CrossRef]
26. Revill, A.S.; Dulvy, N.K.; Holst, R. The survival of discarded lesser-spotted dogfish (*Scyliorhinus canicula*) in the western english channel beam trawl fishery. *Fish. Res.* **2005**, *71*, 121–124. [CrossRef]

27. Baro, J.; Muñoz de los Reyes, I. Bottom trawl fishing yield and selectivity comparisons between square and diamond meshes (Comparación de los rendimientos pesqueros y la selectividad del arte de arrastre empleando mallas cuadradas y rómbicas en el copo). *Informes Tecnicos del Instituto Español de Oceanografia* **2007**, *188*, 1–23.

28. Rodriguez-Cabello, C.; Fernandez, A.; Olaso, I.; Sanchez, F. Survival of Lesser Spotted Dogfish (*Scyliorhinus canicula*, L.) Discarded by Trawlers. *ICES CM 2001/N* **2001**, *6*, 1–10.

29. Frick, L.H.; Walker, T.I.; Reina, R.D. Immediate and delayed effects of gill-net capture on acid-base balance and intramuscular lactate concentration of gummy sharks, *Mustelus antarcticus*. *Comp. Biochem. Physiol. A* **2012**, *162*, 88–93. [CrossRef] [PubMed]

30. Synnes, M.; Larssen, W.E.; Kjerstad, M. Chemical characterization and properties of five deep-sea fish species. *LWT Food Sci. Technol.* **2007**, *40*, 1049–1055. [CrossRef]

31. Costas, B.; Conceicao, L.; Aragao, C.; Martos, J.A.; Ruiz-Jarabo, I.; Mancera, J.; Afonso, A. Physiological responses of Senegalese sole (*Solea senegalensis* Kaup, 1858) after stress challenge: Effects on non-specific immune parameters, plasma free amino acids and energy metabolism. *Aquaculture* **2011**, *316*, 68–76. [CrossRef]

32. Frick, L.H.; Walker, T.I.; Reina, R.D. Trawl capture of Port jackson sharks, *Heterodontus portusjacksoni*, and gummy sharks, *Mustelus antarcticus*, in a controlled setting: Effects of tow duration, air exposure and crowding. *Fish. Res.* **2010**, *106*, 344–350. [CrossRef]

33. Cheuvront, S.N.; Kenefick, R.W. Dehydration: Physiology, assessment, and performance effects. *Comp. Physiol.* **2014**, *4*, 257–285.

34. Herrera, M.; Aragao, C.; Hachero, I.; Ruiz-Jarabo, I.; Vargas-Chacoff, L.; Mancera, J.M.; Conceicao, L. Physiological short-term response to sudden salinity change in the Senegalese sole (*Solea senegalensis*). *Fish Physiol. Biochem.* **2012**, *38*, 1741–1751. [CrossRef] [PubMed]

35. Deck, C.A.; Bockus, A.B.; Seibel, B.A.; Walsh, P.J. Effects of short-term hyper- and hypo-osmotic exposure on the osmoregulatory strategy of unfed north Pacific spiny dogfish (*Squalus suckleyi*). *Comp. Biochem. Physiol. A* **2016**, *193*, 29–35. [CrossRef] [PubMed]

36. Rincón, L.; Castro, P.L.; Álvarez, B.; Hernández, M.D.; Álvarez, A.; Claret, A.; Guerrero, L.; HGinés, R. Differences in proximal and fatty acid profiles, sensory characteristics, texture, colour and muscle cellularity between wild and farmed blackspot seabream (*Pagellus bogaraveo*). *Aquaculture* **2016**, *451*, 195–204. [CrossRef]

37. Moreno, H.M.; Montero, M.P.; Gomez-Guillen, M.C.; Fernandez-Martin, F.; Morkore, T.; Borderias, J. Collagen characteristics of farmed Atlantic salmon with firm and soft fillet texture. *Food Chem.* **2012**, *134*, 678–685. [CrossRef] [PubMed]

38. Slattery, S.L.; Cusack, A.; Nottingham, S.; Bremner, A.P.; Pender, P. Handling of two tropical Australian sharks to improve quality and to identify the cause of tough texture. *J. Aquat. Food Prod. Technol.* **2003**, *12*, 3–28. [CrossRef]

39. Ocaño-Higuera, V.M.; Marquez-Ríos, E.; Canizales-Dávila, M.; Castillo-Yáñez, F.J.; Pacheco-Aguilar, R.; Lugo-Sánchez, M.E.; García-Orozco, K.D.; Graciano-Verdugo, A.Z. Postmortem changes in cazon fish muscle stored on ice. *Food Chem.* **2009**, *116*, 933–938. [CrossRef]

40. Gaarder, M.O.; Bahuaud, D.; Veiseth-Kent, E.; Morkore, T.; Thomassen, M.S. Relevance of calpain and calpastatin activity for texture in super-chilled and ice-stored Atlantic salmon (*Salmo salar* L.) fillets. *Food Chem.* **2012**, *132*, 9–17. [CrossRef] [PubMed]

41. Kiessling, A.; Espe, M.; Ruohonen, K.; Morkore, T. Texture, gaping and colour of fresh and frozen Atlantic salmon flesh as affected by pre-slaughter isoeugenol or $CO_2$ anaesthesia. *Aquaculture* **2004**, *236*, 645–657. [CrossRef]

42. Erikson, U.; Misimi, E. Atlantic salmon skin and fillet color changes effected by perimortem handling stress, rigor mortis, and ice storage. *J. Food Sci.* **2008**, *73*, C50–C59. [CrossRef] [PubMed]

43. Wood, C.M.; Walsh, P.J.; Kajimura, M.; McClelland, G.B.; Chew, S.F. The influence of feeding and fasting on plasma metabolites in the dogfish shark (*Squalus acanthias*). *Comp. Biochem. Physiol. A Mol. Integr. Physiol.* **2010**, *155*, 435–444. [CrossRef] [PubMed]

44. Wood, C.M.; Kajimura, M.; Bucking, C.; Walsh, P.J. Osmoregulation, ionoregulation and acid-base regulation by the gastrointestinal tract after feeding in the elasmobranch (*Squalus acanthias*). *J. Exp. Biol.* **2007**, *210*, 1335–1349. [CrossRef] [PubMed]

45. Murray, C.; Connors, R.; O'Connor, I.; Dowling, V. The physiological response and recovery of a common elasmobranch bycatch species: The lesser spotted dogfish (*Scyliorhinus canicula*) subject to a controlled exposure event. *Biol. Environ. Proc. R. Irish Acad.* **2015**, *115B*, 143–156. [CrossRef]
46. Piiper, J.; Meyer, M.; Drees, F. Hydrogen ion balance in the elasmobranch *Scyliorhinus stellaris* after exhausting activity. *Respir. Physiol.* **1972**, *16*, 290–303. [CrossRef]
47. Frick, L.H.; Reina, R.D.; Walker, T.I. The physiological response of Port Jackson sharks and Australian swellwahrks to sedation, gillnet capture, and repeated sampling in captivity. *N. Am. J. Fish. Manag.* **2009**, *29*, 127–139. [CrossRef]
48. Vargas-Chacoff, L.; Saavedra, E.; Oyarzún, R.M.-M.E.; Pontigo, J.P.; Yáñez, A.; Ruiz-Jarabo, I.; Mancera, J.M.; Ortiz, E.; Bertrán, C. Effects on the metabolism, growth, digestive capacity and osmoregulation of juvenile of sub-antarctic nototheniod fish *Eleginops maclovinus* acclimated at different environmental salinities. *Fish Physiol. Biochem.* **2015**, *41*, 1369–1381. [CrossRef] [PubMed]
49. Foster, C.; Amado, E.M.; Souza, M.M.; Freire, C.A. Do osmoregulators have lower capacity of muscle water regulation than osmoconformers? A study on decapod crustaceans. *J. Exp. Zool. A* **2010**, *313*, 80–94. [CrossRef] [PubMed]

![fishes logo] *fishes*

**MDPI**

*Article*

# Effects of Ortho-Phosphate on Growth Performance, Welfare and Product Quality of Juvenile African Catfish (*Clarias gariepinus*)

**Sebastian M. Strauch** [1,*] ⓘ, **Judith Bahr** [1], **Björn Baßmann** [1] ⓘ, **Adrian A. Bischoff** [1],
**Michael Oster** [2], **Berit Wasenitz** [1] and **Harry W. Palm** [1] ⓘ

[1]  Professorship Aquaculture and Sea-Ranching, Faculty of Agricultural and Environmental Sciences, University of Rostock, Justus-von-Liebig-Weg 6, 18059 Rostock, Germany; judith.bahr@web.de (J.B.); Bjoern.bassmann@gmx.de (B.B.); Adrian.bischoff-lang@uni-rostock.de (A.A.B.); Berit.wasenitz@uni-rostock.de (B.W.); harry.palm@uni-rostock.de (H.W.P.)

[2]  Leibniz Institute for Farm Animal Biology (FBN), Wilhelm-Stahl-Allee 2, 18196 Dummerstorf, Germany; oster@fbn-dummerstorf.de

*   Correspondence: sebastian.strauch@uni-rostock.de

check for updates

Received: 28 November 2018; Accepted: 14 January 2019; Published: 23 January 2019

**Abstract:** Ortho-phosphate inside recirculation aquaculture systems is limited as a consequence of precipitation and regular water exchange rates. To improve plant growth in coupled aquaponics, phosphate fertilizer addition to hydroponics can increase $PO_4^{3-}$-P concentrations inside the process water. We investigated the effects of four $PO_4^{3-}$-P concentrations (<10 (P0), 40, 80, 120 mg $L^{-1}$) in rearing water on growth performance, feed efficiency, and welfare traits of juvenile African catfish (*Clarias gariepinus* Burchell, 1822). By trend, optimum specific growth rate of 2.66% $d^{-1}$ and feed conversion ratio of 0.71 were observed at 40 and 80 mg $L^{-1}$ $PO_4^{3-}$-P. Higher $PO_4^{3-}$-P significantly affected skin coloration, swimming activity and external injuries, with the palest and inactive fish combined with most external injuries in the P120 group. Mineral and protein contents in the fish remained unaffected, while fat content inside the fillets enriched with increasing $PO_4^{3-}$-P. Inorganic P in blood plasma increased significantly, while phosphate concentrations inside the fillet remained unchanged. We suggest that $PO_4^{3-}$-P concentrations of 40 to 80 mg $L^{-1}$ do not reduce the performance of African catfish aquaculture, while increased values of 120 mg $L^{-1}$ affect fish welfare. This allows limited addition of $PO_4^{3-}$-P fertilizer in coupled aquaponics with African catfish to support plant growth.

**Keywords:** African catfish (*Clarias gariepinus*); growth; feed efficiency; elevated phosphate concentrations; welfare

## 1. Introduction

Phosphorus contributes essentially to the growth of animals and plants, because it plays a key role in many vital components [1–3]. In aquaculture systems, fish satisfy their demand for P with their diet [4], while in hydroponic systems, inorganic fertilizer supplies P (as $PO_4^{3-}$-P) to the plants [5,6]. In the combination of both, known as aquaponics, the plants will ideally satisfy their demand from unretained P excreted by the fish [7,8]. However, recent evidence suggests that the levels of P originating from aquaculture process water are low due to rapid precipitation, limiting its use for an optimal plant growth [9–11]. In decoupled aquaponic systems, this deficit can be overcome by addition of inorganic phosphate fertilizer [12,13]. The water used in decoupled aquaponics is not recirculated back to the fish once used in hydroponics, which means that nutrient adjustment cannot affect the fish. In coupled aquaponic systems, the process water circulates between the fish and the plant units and therefore artificial nutrient adjustment in the hydroponics may affect the fish. In the African catfish, as for all other freshwater fish species, the effects of elevated $PO_4^{3-}$-P levels on fish health, welfare and performance have not been studied. In humans, excess levels of inorganic, dietary P resulted in the formation of calcium-phosphate complexes, reducing the levels of serum calcium below optimum levels and causing pathological vascular calcification [14].

The effects of dietary P deficiency on the physiology of teleosts are very well described [4], such as for grouper (*Epinephelus coioides*) [15], European sea bass (*Dicentrarchus labrax*) [16], and silver perch (*Bidyanus bidyanus*) [17]. Deficiency symptoms can be expressed as reduced appetite [15], growth retardation [15,16], reduced feed efficiency [15] and poor bone and scale mineralization [15]. However, the literature lacks information on the effects of increased levels of $PO_4^{3-}$-P on performance and welfare traits in commercial aquaculture species. The acute toxicity of trisodium phosphate ($Na_3 PO_4 12H_2O$) on African catfish (*C. gariepinus*) fingerlings was studied by Ufodike [18], indicating 96 h lethal dose (LD) 50 of 61 mg $PO_4^{3-}$-P $L^{-1}$. In contrast, van Bussel et al. [19] investigated the growth, feed intake, nutrient utilization and health status of juvenile turbot (*Psetta maxima*) reared in intensive recirculating aquaculture systems (RAS). The authors demonstrated that levels of 82 mg $L^{-1}$ of $PO_4^{3-}$-P did not affect the health status of the turbot, and an optimum growth, feed intake and feed efficiency could be achieved at 27 mg $L^{-1}$. In addition, under normal operation of intensive African catfish RAS the levels of $PO_4^{3-}$-P are in the range of 18 to 34 mg $L^{-1}$ [20], without notable mortalities. The causal link of the mortality that was observed previously [18] to $PO_4^{3-}$-P concentrations remain elusive.

Because of the relevance of African catfish in RAS and aquaponic production systems, and the recommendation to elevate $PO_4^{3-}$-P in coupled aquaponic systems for improved plant growth (30–60 mg $L^{-1}$) [21], we assessed the effects of elevated levels of $PO_4^{3-}$-P on performance and welfare traits and other potential effects to be observed for this species. Consequently, the experimental design comprised levels of dissolved $PO_4^{3-}$-P beyond the published $PO_4^{3-}$-P challenges to approximate rearing possibilities in high accumulation aquaculture systems. Implications for a $PO_4^{3-}$-P fertilization in coupled aquaponics systems are discussed.

## 2. Results

### 2.1. Water Quality

On day one of the experiment, the water quality parameters inside the four systems ranged from: dissolved oxygen (DO) = 7.5–7.8 mg $L^{-1}$, temperature = 26.3–27.3 °C, pH-value = 8.5–8.8, conductivity = 560–611 µS $cm^{-1}$, redox-potential = 143–169 mV, $NH_4^+$-N = <0.01–0.04 mg $L^{-1}$, $NO_2^-$-N = 0.01–0.02 mg $L^{-1}$, $NO_3^-$-N = 2–3 mg $L^{-1}$, $PO_4^{3-}$-P = 0.1–0.2 mg $L^{-1}$, $K^+$ = 4–6 mg $L^{-1}$, $Mg^{2+}$ = 9–14 mg $L^{-1}$ and $Ca^{2+}$ = 68–87 mg $L^{-1}$. The mean values and standard deviations of the different water quality parameters over the run of the experiment are given in Table 1.

Significant differences ($p < 0.05$) were revealed for DO, temperature, pH-value, salinity, redox-potential, levels of $PO_4^{3-}$-P, $Ca^{2+}$ and $NO_2^-$-N (Kruskal-Wallis). No significant differences were revealed for conductivity, total dissolved nitrogen (TDN), $NH_4^+$-N and $NO_3^-$-N ($p \geq 0.05$) (Kruskal-Wallis). The DO and the temperature in P120 were lower when compared with the other groups. The highest pH was observed in P0, the lowest in P40 and P80. The highest salinity/conductivity was observed in P120, the lowest in P80. The highest redox-potential was observed in P80 and P120, the lowest in P0. $NH_4^+$-N ranged from 2.19–2.75 mg $L^{-1}$, $NO_2$-N from 0.21–0.34 mg $L^{-1}$. Although not significantly, total oxidized nitrogen (TON), $NO_3^-$-N, TDN and $K^+$ increased by trend from P0 to P120.

For $Ca^{2+}$, significant differences were revealed for all groups, except between P80 and P40, and P80 and P120, with the highest concentrations in P0, and the lowest in P120. For $PO_4^{3-}$-P, significance was revealed for all groups except between groups P80 and P120. $PO_4^{3-}$-P increased from P0 to P80, and by trend to P120. The reference group reached a mean $PO_4^{3-}$-P concentration of 3.6 mg $L^{-1} \pm 2.4$ standard deviation (SD) due to unretained P by the fish.

**Table 1.** Water quality parameters (mean ± standard deviation (SD)).

| Parameter | Unit | P 0 | P40 | P80 | P120 | *p*-Value |
|---|---|---|---|---|---|---|
| | | | | **Group** | | |
| DO | mg L$^{-1}$ | 6.4 $^a$ ± 0.5 | 6.4 $^a$ ± 0.6 | 6.3 $^a$ ± 0.5 | 6.0 $^b$ ± 0.6 | ≤0.001 |
| DO | % | 80 $^a$ ± 6.2 | 80 $^a$ ± 7.3 | 80 $^a$ ± 6.1 | 75 $^b$ ± 7.3 | ≤0.001 |
| Temperature | °C | 27.2 $^a$ ± 0.4 | 27.1 $^a$ ± 0.5 | 27.2 $^a$ ± 0.2 | 26.8 $^b$ ± 0.6 | ≤0.001 |
| pH-value | - | 6.5 $^a$ ± 0.8 | 6.3 $^b$ ± 0.8 | 6.2 $^b$ ± 0.8 | 6.3 $^{ab}$ ± 0.8 | ≤0.001 |
| Salinity | ‰ | 0.50 $^{ab}$ ± 0.11 | 0.50 $^{ab}$ ± 0.11 | 0.49 $^a$ ± 0.11 | 0.52 $^b$ ± 0.15 | 0.044 |
| Conductivity | μS cm$^{-1}$ | 1044 ± 225 | 1056 ± 219 | 1042 ± 228 | 1093 ± 294 | 0.114 |
| Redox potential | mV | 165 $^a$ ± 20 | 181 $^b$ ± 17 | 187 $^c$ ± 14 | 189 $^c$ ± 12 | ≤0.001 |
| NH$_4^+$-N | mg L$^{-1}$ | 2.2 ± 3.6 | 2.8 ± 4.6 | 2.6 ± 4.2 | 2.4 ± 4.2 | 0.252 |
| NO$_2^-$-N | mg L$^{-1}$ | 0.34 $^a$ ± 0.25 | 0.27 $^{ab}$ ± 0.35 | 0.31 $^{ab}$ ± 0.41 | 0.21 $^b$ ± 0.16 | 0.036 |
| TON | mg L$^{-1}$ | 48.4 ± 20.2 | 49.5 ± 20.0 | 53.0 ± 21.1 | 55.2 ± 23.6 | 0.336 |
| NO$_3^-$-N | mg L$^{-1}$ | 48.0 ± 20.1 | 49.2 ± 19.5 | 52.7 ± 21.1 | 55.0 ± 23.7 | 0.309 |
| TDN | mg L$^{-1}$ | 50.6 ± 22.5 | 52.2 ± 22.5 | 55.7 ± 23.9 | 57.5 ± 26.4 | 0.53 |
| PO$_4^{3-}$-P | mg L$^{-1}$ | 3.6 $^a$ ± 2.4 | 39.1 $^b$ ± 20.6 | 62.0 $^c$ ± 30.1 | 103.2 $^c$ ± 54.5 | ≤0.001 |
| K$^+$ | mg L$^{-1}$ | 9 ± 7 | 10 ± 9 | 10 ± 9 | 11 ± 9 | 0.199 |
| Mg$^{2+}$ | mg L$^{-1}$ | 14.9 ± 2.2 | 14.3 ± 2.4 | 13.9 ± 2.9 | 13.7 ± 3.6 | 0.2 |
| Ca$^{2+}$ | mg L$^{-1}$ | 109 $^a$ ± 13 | 94 $^b$ ± 13 | 91 $^{bc}$ ± 15 | 83 $^c$ ± 22 | ≤0.001 |

Superscript letters indicate significant differences between the experimental groups ($p < 0.05$). DO: dissolved oxygen, TON: total oxidized nitrogen (NO$_2^-$-N + NO$_3^-$-N), TDN: total dissolved nitrogen (NH$_4^+$-N + NO$_2^-$-N + NO$_3^-$-N).

With increased addition of PO$_4^{3-}$-P to the three test groups, the water increasingly turned milky. The intensity of milkiness (impression of the observers) reduced over the course of the experiment but remained until the end. On the last day of the experiment, the water quality parameters differed from the conditions of day one and ranged from: DO = 6.2–6.9 mg L$^{-1}$, temperature = 25.7–27.0 °C, pH-value = 4.67–5.94, conductivity = 1099–1151 μS cm$^{-1}$, redox-potential = 165–218 mV, NH$_4^+$-N = 8.38–10.39 mg L$^{-1}$, NO$_2^-$-N = 0.09–0.16 mg L$^{-1}$, NO$_3^-$-N = 81–95 mg L$^{-1}$, PO$_4^{3-}$-P = 7.6–97.2 mg L$^{-1}$, K$^+$ = 16–33 mg L$^{-1}$, Mg$^{2+}$ = 12–16 mg L$^{-1}$ and Ca$^{2+}$ = 101–111 mg L$^{-1}$.

## 2.2. Fish Performance and Body Composition

During the experiment, three animals died in the group P0, two in P80, while no mortality was observed in the groups P40 and P120. After acclimatization, fish always ate the scheduled daily ratio. The means and SD of the fish growth performance parameters are given in Table 2. Final weight, total length, standard length, growth, fillet ratio, specific growth rate (SGR), feed conversion ratio (FCR), and total feed intake (TFI) showed no significant differences between the treatment groups ($p \geq 0.05$).

**Table 2.** Fish growth and feed efficiency of *C. gariepinus* (mean ± SD).

| Parameter | Unit | P0 | P40 | P80 | P120 | *p*-Value * |
|---|---|---|---|---|---|---|
| | | | | **Group** | | |
| Final weight (W$_t$) | g fish$^{-1}$ | 132.9 ± 34.1 | 148.9 ± 51.4 | 161.3 ± 44.0 | 159.6 ± 46.8 | 0.822 |
| Total length | cm | 27.3 ± 2.4 | 28.0 ± 3.3 | 28.9 ± 2.5 | 28.3 ± 2.3 | 0.615 |
| Standard length | cm | 24.4 ± 2.1 | 25.1 ± 3.0 | 25.9 ± 2.3 | 25.4 ± 2.4 | 0.627 |
| Growth (G) | g fish$^{-1}$ | 105.8 ± 3.5 | 118.8 ± 1.1 | 128.5 ± 3.0 | 126.4 ± 4.5 | 0.153 |
| Fillet ratio | % | 33.3 ± 2.3 | 32.3 ± 3.4 | 33.9 ± 1.9 | 33.5 ± 2.2 | 0.155 |
| SGR | % | 2.64 ± 0.05 | 2.66 ± 0.03 | 2.6 ± 0.05 | 2.61 ± 0.03 | 0.172 |
| FCR | - | 0.72 ± 0.05 | 0.71 ± 0.01 | 0.71 ± 0.03 | 0.73 ± 0.02 | 0.450 |
| TFI | g fish$^{-1}$ | 76.0 ± 1.1 | 84.1 ± 1.9 | 91.3 ± 5.2 | 92.8 ± 2.2 | 0.266 |

Superscript letters indicate significant differences between the experimental groups ($p < 0.05$). * *p*-value taken from analysis of variance (ANOVA)/Kruskal-Wallis. SGR: specific growth rate, FCR: feed conversion ratio, TFI: total feed intake.

The mean values of the body composition of the baseline, as well as the control and the experimental groups, are given for whole fish in Table 3 and fillet in Table 4. Except for the fat content in the fillets, no significant differences were found in the whole body and the fillet composition.

The fat content in the fillet was significantly higher in P120 when compared with P0. By trend, the fat content in the fillet and the Ca content in the whole fish increased from group P0 to P120.

**Table 3.** Initial and final body composition of *C. gariepinus* (n = 3).

| Parameter | Unit | Baseline | Group P0 | P40 | P80 | P120 | *p*-Value * |
|---|---|---|---|---|---|---|---|
| Dry matter | g kg$^{-1}$, ww | 207 | 247 ± 13 | 253 ± 1 | 252 ± 8 | 270 ± 8 | 0.146 |
| Ash | g kg$^{-1}$, dm | 131 | 128 ± 10 | 136 ± 8 | 131 ± 5 | 132 ± 6 | 0.780 |
| Protein | g kg$^{-1}$, dm | 694 | 668 ± 13 | 644 ± 18 | 664 ± 17 | 645 ± 2 | 0.275 |
| Fat | g kg$^{-1}$, dm | 97 | 146 ± 19 | 146 ± 17 | 139 ± 11 | 165 ± 10 | 0.401 |
| Calcium | g kg$^{-1}$, dm | 36.8 | 30.2 ± 6.9 | 31.3 ± 4.8 | 33.3 ± 3.1 | 37.1 ± 5.0 | 0.570 |
| Phosphorus | g kg$^{-1}$, dm | 25.3 | 23.7 ± 2.8 | 22.9 ± 1.4 | 24.3 ± 1.1 | 24.4 ± 2.7 | 0.893 |
| Sodium | g kg$^{-1}$, dm | 4.6 | 4.8 ± 0.7 | 3.9 ± 0.7 | 4.5 ± 0.4 | 4.2 ± 0.8 | 0.612 |
| Magnesium | g kg$^{-1}$, dm | 1.5 | 1.7 ± 0.2 | 1.5 ± 0.2 | 1.6 ± 0.2 | 1.4 ± 0.1 | 0.412 |
| Potassium | g kg$^{-1}$, dm | 14.6 | 11.4 ± 0.8 | 12.0 ± 1.3 | 12.8 ± 0.6 | 12.1 ± 0.6 | 0.520 |

Superscript letters indicate significant differences between the experimental groups ($p < 0.05$). *: p-value based on comparison of P0, P40, P80, and P120. *p*-value taken from ANOVA/Kruskal-Wallis. ww: wet weight, dm: dry matter.

**Table 4.** Initial and final fillet composition of *C. gariepinus* (n = 3).

| Parameter | Unit | Baseline | Group P0 | P40 | P80 | P120 | *p*-Value * |
|---|---|---|---|---|---|---|---|
| Dry matter | g kg$^{-1}$, ww | 206 | 222 ± 7 | 222 ± 4 | 222 ± 6 | 216 ± 10 | 0.965 |
| Ash | g kg$^{-1}$, dm | 65 | 58 ± 1 | 57 ± 2 | 55 ± 2 | 55 ± 1 | 0.169 |
| Protein | g kg$^{-1}$, dm | 873 | 836 ± 4 | 851 ± 5 | 827 ± 10 | 836 ± 5 | 0.073 |
| Fat | g kg$^{-1}$, dm | 46 | 81 [a] ± 7 | 101 [ab] ± 8 | 105 [ab] ± 6 | 110 [b] ± 13 | 0.042 |
| Calcium | g kg$^{-1}$, dm | 2.4 | 0.8 ± <0.1 | 0.7 ± 0.1 | 0.8 ± <0.1 | 0.5 ± 0.1 | 0.094 |
| Phosphorus | g kg$^{-1}$, dm | 12.3 | 10.2 ± 0.1 | 10.2 ± 0.4 | 10.3 ± 0.2 | 10.0 ± 0.1 | 0.475 |
| Sodium | g kg$^{-1}$, dm | 2.5 | 2.2 ± 0.2 | 2.0 ± 0.2 | 1.8 ± 0.4 | 1.8 ± 0.1 | 0.725 |
| Magnesium | g kg$^{-1}$, dm | 1.6 | 1.4 ± 0.1 | 1.4 ± <0.1 | 1.3 ± <0.1 | 1.4 ± <0.1 | 0.179 |
| Potassium | g kg$^{-1}$, dm | 21 | 19 ± 2 | 20 ± 2 | 18 ± 1 | 18 ± 1 | 0.458 |

Superscript letters indicate significant differences between the experimental groups ($p < 0.05$). * *p*-value based on comparison of P0, P40, P80, and P120. *p*-value taken from ANOVA/Kruskal-Wallis if $p \geq 0.05$, and from Post-Hoc if *p*-value from ANOVA/Kruskal-Wallis < 0.05.

No significant differences in the values of the apparent net nutrient utilization (ANNU) were found (Table 5). By trend, the ANNUs of Ca and P increased from group P0 to P120, and the ANNU of protein increased by trend slightly from P40 to P120.

**Table 5.** Apparent net nutrient utilization of *C. gariepinus* %.

| Parameter | Group P0 | P40 | P80 | P120 | *p*-Value * |
|---|---|---|---|---|---|
| Protein | 46.3 ± 3.1 | 46.2 ± 1.6 | 47.6 ± 1.6 | 48.3 ± 2.5 | 0.760 |
| Fat | 52.4 ± 11.3 | 53.7 ± 6.8 | 50.7 ± 4.8 | 64.0 ± 4.5 | 0.333 |
| Calcium | 46.2 ± 14.0 | 50.5 ± 10.0 | 54.2 ± 6.8 | 65.0 ± 12.0 | 0.589 |
| Phosphorus | 54.1 ± 8.0 | 54.2 ± 3.7 | 57.8 ± 2.8 | 61.0 ± 9.3 | 0.686 |
| Sodium | 40.5 ± 4.3 | 32.9 ± 7.0 | 39.1 ± 3.9 | 38.1 ± 7.3 | 0.740 |
| Magnesium | 33.2 ± 4.0 | 31.5 ± 4.7 | 32.1 ± 5.4 | 28.7 ± 3.6 | 0.784 |
| Potassium | 38.4 ± 4.4 | 42.4 ± 6.0 | 45.7 ± 2.9 | 44.9 ± 5.1 | 0.455 |

Superscript letters indicate significant differences between the experimental groups ($p < 0.05$). * *p*-value based on comparison of P0, P40, P80, and P120. *p*-value taken from ANOVA/Kruskal-Wallis.

## 2.3. Welfare

The examination of the gills revealed no structural changes of the epithelial tissues, the primary and secondary lamella or the mucous cells (Figure 1).

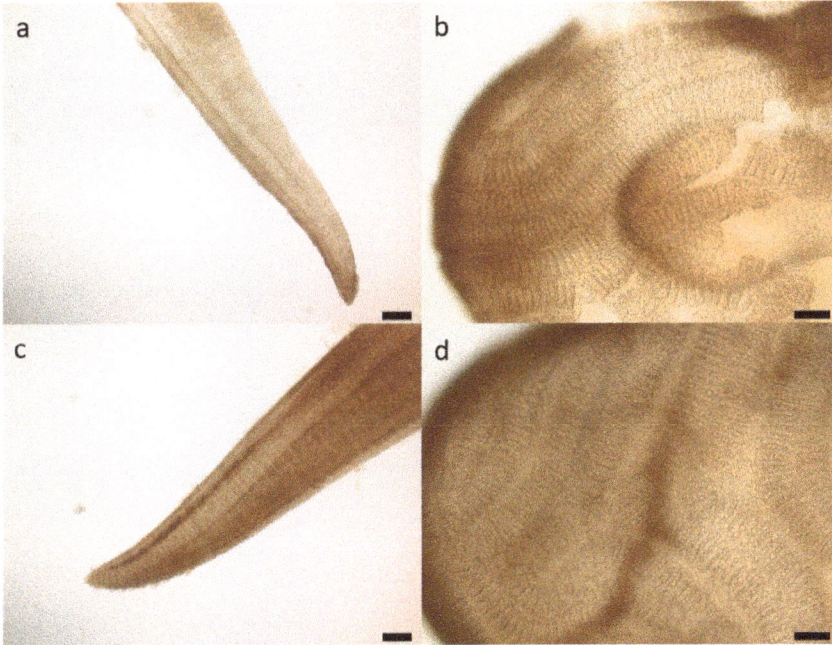

**Figure 1.** Gill filament of the P0 group (**a,b**) and the P120 group (**c,d**) without notable changes, scale bar 200 μm (**a,c**) and 100 μm (**b,d**).

The mean values and SD of the welfare indicators based on the skin coloration, plasma metabolites, and the behavioral analyses are given in Table 6. Significant differences ($p < 0.05$) in skin coloration (Kruskal-Wallis) were observed with the palest fish in P80 and P120 and the darkest in P40 (skin darkness: P40 > P80, P120).

For the functional blood parameters, significant differences ($p < 0.05$) were revealed for plasma concentrations of inorganic P between P120 and P0 and P40 (variance analysis, PROC MIXED). No significant differences were revealed for plasma Ca, $NH_4^+$, glucose and cortisol. Plasma concentrations of inorganic P was increased in animals kept at P120 compared to animals kept at P0 and P40 (P0 ≤ P120; P40 ≤ P120).

The behavioral responses of *C. gariepinus* were slightly altered between the treatment groups (Table 6). Significant differences ($p < 0.05$) were observed within swimming activity (analysis of variance (ANOVA)), air breathing (ANOVA), and the number of biting wounds (Kruskal-Wallis), with lowest swimming activity and air breathing, but most biting wounds in the P120 group. The number of biting wounds was numerically lowest in P0, but no significant difference ($p ≥ 0.05$) (Kruskal-Wallis) was detected between P0 and P80, and P40 and P120. By trend, the observed agonistic behavior (group), swimming activity and air breathing reduced with increasing concentrations of $PO_4^{3-}$-P.

**Table 6.** Health and welfare indicators—plasma metabolites and skin darkness and behavior of *C. gariepinus* challenged with different amounts of dissolved phosphate (means ± SD).

| | | | Plasma Metabolites | | | |
|---|---|---|---|---|---|---|
| | | | | Group | | |
| Parameter | Unit | P0 | P40 | P80 | P120 | *p*-Value |
| Inorganic phosphorus | mg dL$^{-1}$ | 6.22 [a] ± 0.57 | 6.28 [a] ± 0.94 | 6.68 [ab] ± 0.69 | 7.17 [b] ± 1.14 | 0.028 |
| Calcium | mg dL$^{-1}$ | 14.27 ± 1.30 | 14.44 ± 1.09 | 14.68 ± 1.13 | 14.39 ± 0.92 | 0.740 |
| Ammonia | µg dL$^{-1}$ | 409 ± 172 | 400 ± 234 | 338 ± 79 | 344 ± 91 | 0.514 |
| Blood glucose | mmol L$^{-1}$ | 2.34 ± 0.39 | 2.21 ± 0.43 | 2.44 ± 0.58 | 2.35 ± 0.42 | 0.492 |
| Plasma cortisol | ng mL$^{-1}$ | 28.5 ± 13.3 | 26.9 ± 19.9 | 26.4 ± 10.3 | 29.0 ± 17.0 | 0.716 |
| Skin darkness | - | 6.2 [ab] ± 1.6 | 7.2 [a] ± 1.1 | 5.5 [b] ± 1.7 | 5.3 [b] ± 1.8 | <0.001 |
| | | | Behavior | | | |
| Biting wounds | *n* | 2.6 [a] ± 2.5 | 4.5 [b] ± 3.7 | 4.0 [ab] ± 3.6 | 5.4 [b] ± 4.3 | 0.002 |
| Swimming activity (group) | % | 61.5 [a] ± 22.0 | 53.5 [ab] ± 20.2 | 56.4 [ab] ± 24.5 | 44.4 [b] ± 29.9 | 0.002 |
| Air breathing (group) | *n* | 19.7 [a] ± 5.4 | 15.2 [ab] ± 9.4 | 13.5 [ab] ± 9.3 | 5.2 [b] ± 2.6 | 0.003 |
| Agonistic behavior (group) | *n* | 13.7 ± 9.9 | 11.3 ± 5.7 | 8.5 ± 4.5 | 5.8 ± 3.3 | 0.164 |
| Swimming activity (individual) | % | 62.6 ± 17.3 | 55.7 ± 22.9 | 50.3 ± 15.5 | 43.6 ± 11.2 | 0.383 |
| Air breathing (individual) | % | 3.7 ± 1.4 | 2.3 ± 0.8 | 2.3 ± 2.3 | 1.8 ± 1.5 | 0.197 |
| Agonistic behavior (individual) | *n* | 2.8 ± 1.7 | 3.3 ± 2.1 | 2.2 ± 1.3 | 2.7 ± 2.8 | 0.541 |

Superscript letters indicate significant differences between the experimental groups (*p* < 0.05).

## 3. Discussion

This is the first time that effects of elevated concentrations of $PO_4^{3-}$-P in the rearing water were studied on the growth performance and welfare of African catfish. Medium increase in $PO_4^{3-}$-P concentrations were without significant effects, while the level of 120 mg L$^{-1}$ $PO_4^{3-}$-P affected behavior and welfare.

### 3.1. Water Quality

At the beginning of the experiment (d1), the water quality was similar between the treatment groups, with high DO and pH-values, low conductivity, high redox potential and very low levels of $NH_4^+$-N, $NO_2^-$-N $NO_3^-$-N, and $PO_4^{3-}$-P, originating from the properties of the aerated tap water. Although the pH dropped over the course of the experiment, this change took place during the first 20 days in a continuing manner, allowing the fish to acclimatize to these changing conditions.

Addition of $P_2O_5$ did not simply result in proportional increase of $PO_4^{3-}$-P concentrations inside the water. This disproportional response was probably a result of precipitation of $PO_4^{3-}$-P with $Ca^{2+}$ and $Mg^{2+}$ a phenomenon which is well understood [22,23]. This is supported by the increasing turbidity of the process water, significantly decreasing levels of dissolved $Ca^{2+}$ (*p* < 0.05), and by the trend of decreasing levels of dissolved $Mg^{2+}$ after addition of $P_2O_5$. Using a protocol based on $P_2O_5$ dose and analytical response, the target levels of P40 and P80 were achieved. By trend, the levels of $PO_4^{3-}$-P in the group P120 mostly remained above the other concentrations with a high standard deviation, while the P0 group in maximum reached 8.9 mg L$^{-1}$ caused by natural excretion of the fish.

Only minor differences in the water quality parameters were observed between the groups. $K^+$, $NH_4^+$-N and to lesser extent $NO_2^-$-N started to accumulate from day 30, in all systems, indicating an overload of the biofilters. As regular water exchange remained, the water quality parameters were well appropriate for *C. gariepinus* [24–27]. Given that our statistical analysis revealed significant differences in DO, temperature, pH-value and redox-potential between the groups, we cannot rule out minor influence on our results. However, due to the small magnitudes of these values, and because of the inevitable precipitation of Ca/Mg-phosphates, we estimate the observed effects to be monofactorial, resulting from the major differences in $PO_4^{3-}$-P.

*3.2. Mortality and Growth Performance*

The mortality during the experiment was very low (2.6%). Considering that of five dead fish three died in P0 and two in P80, the mortality was likely not a result of the increased $PO_4^{3-}$-P concentrations in the rearing water.

For juvenile African catfish, we presumed an exponential growth (body weight$_{d0}$ × exp$^{(k \times t)}$, with k = specific growth rate and t = days of the experiment). From our own previously recorded experimental data [20], we presumed a FCR of 1.0 and a daily feed ratio of 2.2. The FCR turned out to be lower (0.71–0.73), which can be considered as normal for smaller/younger fish [11,28]. After the acclimatization period, the fish always ate their scheduled ratio, but appeared to be close to maximum feed intake, which is important in commercial systems to maximize growth. The daily feed ratio was therefore also comparable with that of commercial systems, so we estimated it to be appropriate.

The dietary composition of the commercial feed, especially the levels of P, was adequate according to the literature [4]. Although feed intake appeared to increase with elevated levels of $PO_4^{3-}$-P, this was not a result of increased appetite but in correspondence with the feeding curve and based on the initial body weight. Using the initial life weight as co-variant, no significant differences in TFI were observed. Consequently, optimum FCR (0.71) and SGR (2.66% d$^{-1}$) were by trend best in P40 and P80. Similar positive effects in response to moderately increased concentrations of $PO_4^{3-}$-P were also observed with turbot (*Psetta maxima*) [19] and tilapia (*Oreochromis niloticus*) [29].

The analysis of the whole fish revealed that body composition was in line with literature data, with only the levels of P, Ca and protein being higher, and the largest difference in direct comparison in the group P120 [30–32]. The analysis of the fillets revealed a higher content of ash and protein in our study, a lower fat content, and minerals in a similar range [33–35]. In our study, the analyzed fillets revealed higher proportions of ash, partially higher or comparable protein levels and similar or lower fat contents [31,32,34,35]. The levels of mineral contents were in range with literature data or even higher [31–33]. For commercially produced African catfish (*C. gariepinus*) fillet from intensive production in Germany, Wasenitz et al. [36] found lower levels for ash and $P_2O_5$ 4.2 g kg$^{-1}$ wet weight (ww) (value converted into P was 1.84 g kg$^{-1}$, in ww) and higher values for dry matter and protein. Wasenitz et al. [36] determined almost double values for fat when compared to our data.

An interesting observation is the difference in P and Ca content between the juvenile fish at the beginning of the experiment (baseline) and the final body composition. The juvenile fish were fed with starter feed, containing higher levels of P, protein and lower levels of fat. The young fish had higher levels of Ca and P when compared with the final body composition of the fish at the end of the experiment, which was reduced by about 18% (Ca) in P0. They had neither access to a carnivorous natural diet with higher P/Ca contents when compared with the commercial growth out feed, soluble P, nor to sediments with organic/inorganic P deposits like fish in natural habitats. With increasing $PO_4^{3-}$-P concentrations inside the water, the Ca/P content in the body also increased. This indicates that the P-content in the water influences the mineralization status of the fish bones, where most Ca/P is deposited. We also observed that, if $PO_4^{3-}$-P increases, the fat content increases. This suggests that the fat content in the fillet can be manipulated via the levels of $PO_4^{3-}$-P in the water. Our data demonstrate that the fish is able to uptake $PO_4^{3-}$-P directly from the water, and requires adequate P concentrations to build up fat. The fish held in the P0 group with $PO_4^{3-}$-P < 2.6 mg L$^{-1}$ most likely mobilized P from the bones and were therefore unable to keep up with the growth when compared with the fish in P40 and P80. As a consequence, generally elevated levels of $PO_4^{3-}$-P between 40–80 mg L$^{-1}$ inside the rearing water increase the growth potential of African catfish RAS, because of higher feed efficiency. The content of P inside the fillets was rather constant at about 1% on a dry matter (dm) basis.

When challenged with elevated levels of $PO_4^{3-}$-P, African catfish increased the ANNUs of Ca and P by trend up to 120 mg $L^{-1}$, indicating higher bone mineralization [15]. Likewise, the ANNU of protein slightly increased by trend from group P40 to P120. Table 5 shows that the ANNU-efficiencies of Ca and P increase under elevated levels of $PO_4^{3-}$-P. Because the feed was the same in all groups, the difference of P between P0 and the other groups is a result of utilization of $PO_4^{3-}$-P from the water, allowing better mineralization of the bones, which is also evident in the higher utilization efficiency of Ca under elevated $PO_4^{3-}$-P. This effect obviously occurs under the use of P limited diets.

Growth can be defined as the assimilation of energy in the form of fat and protein. To be available for growth, the gross fat and protein from feed (net energy) must be: (1) digestible; (2) metabolizable; and (3) not required for maintenance energy expenditure [2,37]. Maintenance energy expenditure is the sum of energy used for swimming activity and the maintenance metabolism, the vital life functions [38]. The partition of energy allocated to maintenance is substantial, as it accounts for about 15–30% of the gross energy intake [39]. If growth is increased under restrictive feeding (our impression was that fish were not fully satiated), the observed trend in the P120 must result from reduced maintenance energy expenditure, resulting from elevated levels of $PO_4^{3-}$-P. We observed that the catfish significantly reduced their activity under increased concentrations of $PO_4^{3-}$-P (further discussed in Section 3.3), possibly allowing the fish to utilize feed derived energy and nutrients for growth.

In fish, the maintenance metabolism is mainly dominated by ionic and osmotic regulation [40]. Also the maintenance of phosphorus requires energy. If dietary P is deficient, increased effort (energy) has to be invested to maintain P-homeostasis. Fish are able to uptake P from the water [41]. The primary and subsequent uptake pathways for organic or inorganic P in African catfish were not assessed, however, in trout these were identified in the proximal and distal intestine, subordinate uptake is provided by the gills [42,43]. Reabsorption and excretion of P from the urine is provided through the kidneys [44]. It can be speculated that additional supply of P to the water allowed the animal to reduce the energy expenditure for P uptake from feed and water via intestine and the gills, and possibly for renal resorption. This energy could be incorporated as fat.

Considering that an increase in $PO_4^{3-}$-P also slightly increased protein assimilation efficiency (from P40 to P120), the elevated concentrations of body fat content under high $PO_4^{3-}$-P are unlikely a result of increased lipogenesis of dietary protein, but rather a direct deposition of dietary fat. Consequently, either the concentrations or the digestibility of the dietary P was too low (the feed specifications indicate only plant and animal P sources), or that the uptake via the intestine was exhausted. In either case, $PO_4^{3-}$-P will be used to supplement the needs of the animal. It is conceivable that the variation of $PO_4^{3-}$-P in the water prompted intrinsic responses of major players of P homeostasis. Indeed, studies in fish and mammals showed that the dietary P supply impacts on various tissue sites and body compartments such as gut, kidney and bone [45–47]. However, molecular specificities of associated regulators, transporters, and endocrine and paracrine signals are largely unknown and yet to be elucidated in African catfish. It should be noted that levels of $NH_4^+$ and glucose remained unaffected by treatments which indicate regular nutrient utilization and energy homeostasis. Serum $Ca^{2+}$ and $NH_4^+$ remained unaffected by the treatment, therefore the homeostasis/excretion via the kidney was likely not impaired [25,44,48].

### 3.3. Welfare

The welfare of fish is usually studied via functional and behavioral approaches. Functional approaches are used to evaluate whether an animal can cope physiologically with its environmental or husbandry conditions. Behavioral approaches are used to evaluate the ability to perform natural behavioral patterns [49]. Environmental influences, primarily changes in water quality, can adversely affect fish welfare [50–52]. Thus, it is possible that the physiology, in particular stress response reactions, is influenced [53,54]. Accordingly, behavioral changes may also occur.

We observed that with increasing concentrations of $PO_4^{3-}$-P, African catfish shifted towards a lighter grey tone of the skin. Catfish are able to alter their skin coloration to lighter or darker.

Melanosomes inside the skin are responsible for this action, as they may change in number and size [55,56]. Different mechanisms causing this change are described, such as adaptation to the background color [57,58] and light regime [59] of their environment, dietary composition [60], hormones [55] and environmental stress [61]. A direct effect of $PO_4^{3-}$-P in the water on the skin coloration of fish is not described in literature. However, the skin of the scaleless African catfish could also be affected in regard to mineralization [15]. It is known that the mineralization of the skin can be affected by dietary P [62]. In this case, the lighter skin coloration would be the result of calcium-phosphates that precipitated inside the skin [17,63], to be tested in subsequent studies.

Alterations of the cortisol level is often used as an indicator for stress in welfare investigations [43], because it reflects acute and long-term (chronic) stress [53,64]. Cortisol production can be influenced by environmental impacts, such as pollutants in the water [54]. To our knowledge, there is no study addressing cortisol changes resulting from increased $PO_4^{3-}$-P concentrations in the rearing water available. In our study, we did not find significant differences in the plasma cortisol response between the treatment groups. Slightly elevated plasma cortisol levels due to netting stress occurred over the course of the sampling procedure. Under this consideration, the plasma cortisol amplitude increased regularly with ongoing sampling in every group but was not affected by the $PO_4^{3-}$-P concentration in the rearing water. This also demonstrated that an alteration of the cortisol level due to $PO_4^{3-}$-P concentration did not occur. Alterations of the blood glucose concentrations were similar in all groups and in a typical range for *C. gariepinus* [65]. Further effects such as immuno- and growth-depression as a consequence of chronic hypercortisolism [44] could not be observed in our study. Therefore, we suggest that the stress level following $PO_4^{3-}$-P exposure below 80 mg $L^{-1}$ can be considered as low.

In the P120 group, the fish altered their agonistic behavior, also seen in the highest number of biting wounds compared with P0 to P80. However, differences between P40, P80 and P120, or P0 and P80 were not significant. This trend correlated with significantly less group air breathing and swimming activity from P0 towards P120, leading to the assumption that the observed elevation in the number of biting wounds was due to behavioral changes. One explanation is an adaption to the water conditions, in particular regarding turbidity. African catfish have only poorly developed eyes [66] and are nocturnal. It can be speculated, that inside the turbid water, the fish react differently on each other; therefore interaction (visual, tactile sense) only occurs when the animals are very close, triggering a more aggressive response, such as biting (discussed in Baßmann et al. [65]). African catfish are normally territorial animals. Under high stocking density, this territorial behavior is suppressed, and the fish tend to school up and spend much time resting, provided that feed and DO are adequate. Under high concentrations of DO, the African catfish is described as a facultatative air breather [67]. If however, the fish returns from air breathing at the surface, finding a resting place may disrupt the school and trigger an aggressive response in other fish. Because African catfish are more active at twilight and/or night, Britz and Pienaar [68] demonstrated a link between light intensity, swimming activity, air breathing and agonistic behavior, resulting in injuries and cannibalism. When exposed to bright light, the animals in these studies were less active and reduced air breathing, but spend more time resting, were more territorial and had increased numbers of injuries. This is corresponding to our results. Possibly, the diffuse light conditions during the day in our experiments in combination with the increased water turbidity under high $PO_4^{3-}$-P is perceived brighter than without turbidity by the fish. To find out which exact mechanism (e.g., the visual or tactile sense) caused the increased number in biting wounds under high $PO_4^{3-}$-P concentrations, the experiment would have to be repeated under total darkness.

## 4. Materials and Methods

The experiment and the analyses were performed according to the experimental plan and in accordance with the guidelines and legislation in force, such as the Protection of Animals Act, and the permission for keeping fish under experimental conditions. The study was approved by the 'Landesamt für Landwirtschaft, Lebensmittelsicherheit und Fischerei Mecklenburg-Vorpommern–Veterinärdienste und Landwirtschaft' Rostock, Germany with the following ID number: 7221.3-2-022/16.

### 4.1. Experimental Design

The experimental setup consisted of four separate RAS, located at the aquaculture research facilities at the University of Rostock. Each system included three glass aquaria (100 cm × 40 cm × 50 cm; $volume_{max}$ = 200 L, $volume_{effective}$ = 150 L) and a sump ($volume_{effective}$ = 77 L), resulting in a total, effective volume of 527 L per system. The sump housed a small mechanical filter for the removal of particulate matter (filter pad, 30 ppi), a moving bed biofilter supporting nitrification (carrier material: HEL-X® Biocarrier, 780 g, 836 $m^{-2}$ $m^{-3}$, 165 kg $m^{-3}$ ≙ 3.95 $m^2$ effective surface area), a submerged pump (AquaMedic Ocean Runner OR 3500), which was set to a flow rate of 13 L $min^{-1}$ ≙ 4.3 L $min^{-1}$ per aquarium and a heater (GroTech Heater HI-300, 300 W). Aeration was achieved with a diaphragm pump (1 $RAS^{-1}$, EHEIM air pump 100), which was connected to one air outlet in each aquarium. Each RAS represented an experimental unit only differing in P concentrations (Section 4.3). All RAS were filled with temperature conditioned tap water.

### 4.2. Fish and Feeding

Each RAS was stocked with juveniles of African catfish (16 $tank^{-1}$) with a mean weight (± SD) of 30.81 g ± 6.56 (Fischzucht Abtshagen, Abtshagen, Germany). Prior to the start of the experiment, fish were acclimated to the rearing conditions for seven days. A set of 16 animals was stocked in a separate aquarium as a reference for initial fillet and whole-body composition (baseline). This baseline was treated the same way as the control and the test groups during the acclimatization period (first 7 days), held without feeding for two days and was then slaughtered for analysis.

The fish were hand-fed during the whole experiment (61 d, from stocking to slaughter), solely with a commercial diet for African catfish (Skretting Meerval START, 3.5 mm) at 9:30 am. The feed composition was analyzed with 94.40% dry matter content, 51.30% crude protein, 10.80% fat, 9.85% ash, 2.03% fiber, 2.24% calcium, 0.18% magnesium, 1.55% phosphorus, 1.01% potassium, 0.42% sodium, and was further specified with 5000 UI $kg^{-1}$ retinol (E672), 750 UI $kg^{-1}$ cholecalciferol (E671), 42.0 mg $kg^{-1}$ iron, 2.1 mg $kg^{-1}$ iodine, 5.0 mg $kg^{-1}$ copper, 16.0 mg $kg^{-1}$ manganese, 110.0 mg $kg^{-1}$ zinc, 50.0 mg $kg^{-1}$ ethoxyquin (E324). During acclimatization, all fish were fed 2.0% of body weight $d^{-1}$. The feeding was observed to prevent excess feeding and possible leftovers. Uneaten pellets were carefully collected with a small net after 30 min, counted and multiplied with the dry weight of 0.12 g $pellet^{-1}$. Due to a good feed intake, the feed ratio was gradually increased to reach a mean daily feeding ratio of 2.64% BW $d^{-1}$. Mortality was determined visually. Dead fish were removed from the tank and the mean calculated weight was used to recalculate the feed ratio per fish tank. Before the fish were slaughtered for the analysis of the initial and final body composition, both the fish of the baseline and the experimental groups were held without feeding for two days.

### 4.3. Experimental Units and Water Quality

The concentrations of $PO_4{}^{3-}$-P were aimed at four levels, with group 1 (control: P0, without addition of fertilizer): $PO_4{}^{3-}$-P <10 mg $L^{-1}$, group 2 (P40): 40 mg $L^{-1}$, group 3 (P80): 80 mg $L^{-1}$, and group 4 (P120): 120 mg $L^{-1}$. P addition was achieved by solving diphosphorus pentoxide ($P_2O_5$, Roth, Karlsruhe, Germany) in a mixing tank (HD-polyethylene) in demineralized water. $P_2O_5$ reacts strongly exothermic with water to orthophosphoric acid ($P_4O_{10}$ + 6$H_2O$ → 4$H_3PO_4$). The acidic solution (pH < 2) was then neutralized with sodium hydroxide (NaOH, Roth, Karlsruhe, Germany) to pH 6.5–7.0 and left for temperature conditioning overnight. The neutralized, temperature conditioned $PO_4{}^{3-}$-P solution was then added to the system water into the sump during water exchange. Due to the fact that at neutral to alkaline pH, phosphoric acid precipitates with calcium (3 $CaCO_3$ + 2 $H_3PO_4$ = $Ca_3(PO_4)_2$ + 3 $CO_2$ + 3 $H_2O$) or magnesium (2 $H_3PO_4$ + 3 $MgCO_3$ → $Mg_3$ ($PO_4)_2$ + 3 $H_2O$ + 3 $CO_2$), which are dissolved in tap water, the target concentrations were reached after multiple additions of neutralized $PO_4{}^{3-}$-P solution during the first week of the experiment. To adjust the conductivity of the four treatment groups, sodium chloride was added to the solutions. The general rearing conditions, defined by the levels of DO, pH-value, electrical conductivity and redox-potential, were measured in all tanks in triplicate daily between 7:00 and 9:00 a.m. with a multimeter (HACH® Multimeter HQ40d). To determine the levels of $NH_4{}^+$-N, $NO_2{}^-$-N, $NO_3{}^-$-N, $PO_4{}^{3-}$-P, $K^+$, $Ca^{2+}$ and $Mg^{2+}$, water samples were taken on Mondays, Wednesdays and Fridays at 9:00 am, and were measured colorimetrically with an automated discrete analyzer (ThermoFisher Scientific™ Gallery™) according to the manufacturer's protocol. To measure the effect of water exchange and $PO_4{}^{3-}$-P dosing to the process water, another water sample was taken 24 h after water exchange. All samples were measured in triplicates. To maintain an adequate water quality, water exchange of 75 L $\triangleq$ 15% RAS volume was performed by emptying the pump sumps on Mondays, Wednesdays and Fridays, three hours after feeding.

### 4.4. Water Quality Calculations

Ammonia nitrogen:

$$c_i\left(NH_4^+ - N\right) = \frac{c_i\left(NH_4^+\right)}{M\left(NH_4^+\right)} * M(N) \tag{1}$$

with: $c_i$ = concentration, $M$ = molar mass, $N$ = nitrogen, $NH_4^+$ = ammonia.
Total oxidized nitrogen (*TON*):

$$c_i(TON) = \frac{c_i\left(TON, \text{ as } NO_3^- - N\right)}{M\left(NO_3^-\right)} * M(N) \tag{2}$$

with $c_i$ = concentration, $M$ = molar mass, $N$ = nitrogen, $NO_3^-$ = nitrate, $TON$ = total oxidized nitrogen.
Nitrite nitrogen:

$$c_i\left(NO_2^- - N\right) = \frac{c_i\left(NO_2^-\right)}{M\left(NO_2^-\right)} * M(N) \tag{3}$$

with $c_i$ = concentration, $M$ = molar mass, $N$ = nitrogen, $NO_2^-$ = nitrite.
Nitrate nitrogen:

$$c_i\left(NO_3^- - N\right) = c_i(TON) - c_i\left(NO_2^- - N\right) \tag{4}$$

with $c_i$ = concentration, $N$ = nitrogen, $NO_2^-$ = nitrite, $NO_3^-$ = nitrate, $TON$ = total oxidized nitrogen.
Total dissolved nitrogen (*TDN*):

$$c_i(TDN) = c_i\ TON + c_i\left(NH_4^+ - N\right) \tag{5}$$

with $c_i$ = concentration, $N$ = nitrogen, $TDN$ = total dissolved nitrogen, $TON$ = total oxidized nitrogen.
Ortho-phosphate phosphorus:

$$c_i\left(PO_4^{3-} - P\right) = \frac{\left(PO_4^{3-}\right)}{M\left(PO_4^{3-}\right)} * M(P) \tag{6}$$

with $c_i$ = concentration, $M$ = molar mass, $P$ = phosphorus, $PO_4{}^{3-}$ = ortho-phosphate

### 4.5. Analysis of Feed and Fish

To assess the feed composition and the related growth performance of fish as the overall assimilation efficiency of protein, fat and minerals, feed samples were analyzed for dm content, protein, fat, ash, Ca, total phosphorus (TP), Na, Mg and K at the "Landwirtschaftliche Untersuchungs- und Forschungsanstalt der LMS Agrarberatung GmbH (LUFA) in Rostock, Germany. The feed parameters were measured according to standard methods "VDLUFA", (http://www.vdlufa.de/Methodenbuch/index.php/de/). Total weight, standard length (nose to base of fin), total length (nose to tip of fin), external injuries of all fish were recorded at the beginning and end of the experiment, and skin coloration of all fish was recorded once at the end of the experiment. To determine the skin coloration a numbered grey-chart with nine different shades of grey was used. The fish were placed on the grey chart (RGB codes: 242, 242, 242; 217, 217, 217; 191, 191, 191; 166, 166, 166; 128, 128, 128; 89, 89, 89; 64, 64, 64; 38, 38, 38; 13, 13, 13) immediately after catch, and their individual skin coloration was evaluated according to the best match.

The first eight, randomly caught fish (including 3 males and 3 females) from each tank were used for blood withdrawal and dissection. Prior to blood withdrawal, fish were stunned by percussion of the brain, afterwards they were immediately decapitated. To determine the concentrations of functional blood parameters (plasma cortisol, P, $Ca^{2+}$ and $NH_4^+$ cortisol and glucose), approx. 0.5 mL blood was taken from the *Vena caudalis* of three males and three females of each tank. The blood glucose was immediately measured (ACCU-CHEK® Aviva) after blood withdrawal. The remaining blood was transferred into heparin-coated tubes (Sarstedt Monovette, Li-Heparin LH, 7.5 mL) and stored on ice. Plasma was prepared, and samples were stored at $-20\,^\circ$C until analysis of plasma cortisol, P, $Ca^{2+}$ and $NH_4^+$ concentrations. To measure cortisol, an enzyme-linked immunosorbent assay (ELISA) for cortisol in fish (Cusabio®) with no significant cross-reactivity or interference between fish cortisol and analogues was used according to the manufacturer's protocol. Plasma minerals (inorganic P, $Ca^{2+}$, $NH_4^+$) were analyzed with commercial assays using Fuji DriChem 4000i (FujiFilm, Minato, Japan).

To determine the fillet and organ indices, the fillets, carcass, gonads, hearts, livers, kidneys and gills were weighed. The morphological analysis of the gills was based on Zayed and Mohamed [69]. To determine negative impact on the gill morphology, samples of the gills were taken at the end of the experiment. From each experimental group, the gill arches from five males and five females ($n = 4 \times 10$) were prepared and documented via photography. The pictures of the gill arches were analyzed to determine potential effects caused by differences of rearing conditions. Also, three samples of gill arches were taken from each group for a microscopic assessment. Therefore, gill lamellae from the second gill arch were collected, left in glycerin and 70% ethanol for 24 h at 38 °C, and consequently embedded in paraffin. Glycerin has the same refractive index as cytoplasm, replaces it in the cells, and is therefore useful as mounting medium for light microscopy. Samples of the suprabranchial organ were collected and prepared for microscopic analysis the same way. In case of abnormalities, findings would be compared with the literature.

The fillets of the dissected fish were homogenized three times with a meat mincing machine (BOSCH, MFW 67440) and conveyed for analysis of the same parameters as the feed to the LUFA GmbH (see above). The remaining eight fish from each tank were analyzed as whole fish, where total weight, standard and total length, skin coloration, sex and external injuries were also recorded. These fish were homogenized and analyzed for whole body composition (LUFA GmbH).

### 4.6. Performance Calculations and Definitions

From the data referring to the fish counting, feed input, weight measurements, TFI, growth (G), FCR, SGR, and organ indices were calculated.

Total feed intake (TFI):

$$\text{Feed eaten by fish over the experimental period } \left(\text{g fish}^{-1}\right) \tag{7}$$

Growth (G) (g):

$$G = W_t - W_0 \tag{8}$$

with $W_0$ = initial fish weight, $W_t$ = final fish weight at harvest (last day of experiment).

Feed conversion ratio (FCR):

$$FCR = \frac{TFI}{W_t - W_0} \tag{9}$$

with *TFI* = total feed intake, $W_0$ = initial fish weight, $W_t$ = final fish weight.

Specific growth rate (SGR):

$$SGR \left[ \% \, d^{-1} \right] = \frac{(Ln\,(W_t) - Ln\,(W_0))}{d} * 100 \tag{10}$$

$W_0$ = initial fish weight, $W_t$ = final fish weight, and $d$ = days.

Organ index (OI):

$$OI = \frac{Organ\ weight\ [g]}{Body\ weight\ [g]} * 100 \tag{11}$$

Mortality (Mo):

$$Mo\ [\%] = \frac{Number\ of\ dead\ fish}{Initial\ number\ of\ fish} * 100 \tag{12}$$

Apparent Net Nutrient Utilization (ANNU):

$$ANNU = \frac{(W_t * X_t - W_0 * X_0)}{(TFI * X_F)} * 100 \tag{13}$$

$W_0$ = initial fish weight, $W_t$ = final fish weight, $X_0$ = initial nutrient concentration of the fish, $X_t$ = final nutrient concentration of the fish, $X_F$ = nutrient concentration of the feed.

### 4.7. Fish Behavior

The ethological investigations were based on van de Nieuwegiessen et al. [53]. The behavior was observed in individual animals, as well as in the whole group after the third and fifth week of the experiment. Observation of an individual was as follows: in each tank, one animal was observed for five minutes. Behavioral patterns such as swimming, air breathing and agonistic behavior were quantitatively documented. The ethogram is given in Table 7.

**Table 7.** Ethogram—behavioral patterns and their definition, adapted from van de Nieuwegiessen et al. [53].

| Behavior | Definition |
|---|---|
| Agonistic behavior | Chasing or biting a fish, or being chased upon or bitten by another fish |
| Air breathing | The animal moves to the water surface and takes a gulp of air. This was checked by escaping air from the gills of the fish, when it was swimming back to the bottom of the tank |
| Swimming | A displacement of the body, while browsing, moving, eating and air-breathing |

In case the investigated fish got out of sight during observation, another fish was followed, starting from the position the former was lost and the observation continued according to van de Nieuwegiessen et al. [53]. The group behavior was determined by evaluation of video recordings (from outside the tanks) of five minutes. Thereby, the number of swimming fish was determined in 30 s intervals. The activity pattern was calculated as percentage of swimming in comparison to the total number of fishes. Air breathing and agonistic behavior were documented as frequency.

### 4.8. Data Analyses

Data analyses were performed with SPSS (IBM® SPSS® Statistics Version 22). All data were tested for normal distribution using the Shapiro-Wilk test. When data showed normal distribution, One-Way-ANOVA was performed and homogeneity of variance was tested using the Levene's test: concentrations of $Ca^{2+}$ in the rearing water, initial body weight, whole body contents of ash, protein, fat, Ca, P, Na and K, fat content in fillet, ANNUs of protein, fat, P, Mg and K, heart index, blood glucose, swimming activity (individual), agonistic behavior (individual), air breathing (group).

When data were not normally distributed, the Kruskal-Wallis test was performed: levels of DO, temperature, pH-value, salinity, conductivity, redox-potential and concentrations of $NH_4^+$-N, $NO_2^-$-N, TON, $NO_3^-$-N, TDN, $PO_4^{3-}$-P, $K^+$, $Mg^{2+}$ in the rearing water, whole body contents of dm and Mg, fillet contents of dm, ash, protein, Ca, P, Mg, Na and K, ANNUs of Ca and Na, skin darkness, serum cortisol, biting wounds, swimming activity (group), agonistic behavior (group), air breathing (individual). When One-Way-ANOVA showed significant differences and data showed homogeneity of variances Tukey-HSD (Honestly Significant Difference) was used as post-test (initial body weight, fat content in fillet), when data showed no homogeneity of variances Dunnett-T3 was used as post-hoc-tests (air breathing group).

Plasma P, Ca, $NH_3$ were analyzed via variance analyses (PROC MIXED; SAS version 9.4; SAS Institute, Cary, NC, USA), including effects represented by treatment, tank and sex. Body weight related traits (total length, standard length, fillet ratio, SGR, FCR, TFI) were corrected for initial live weight (mean per tank) used as co-variable. The level of significance was set at $p < 0.05$.

## 5. Conclusions

Economical, ecological and ethical considerations require healthy animals with high feed efficiency in aquaculture production. Modern integrated systems such as aquaponics demand adjustment of water parameters in order to improve plant production, however, with no negative consequences on fish growth, welfare and product quality. This study demonstrates that different $PO_4{}^{3-}$-P levels between 3.6 and 120 mg $L^{-1}$ do not affect the P contents of 1% inside the fillet of African catfish. However, the fat content significantly increases with increasing $PO_4{}^{3-}$-P, with relevance for product quality. The near absence of $PO_4{}^{3-}$-P inside the water, combined with the absence of sediments as P deposits, results in a trend of reduced growth, indicating the importance of dissolved or precipitated P availability inside the system even under the use of commercial diets. Elevated levels of $PO_4{}^{3-}$-P up to 80 mg $L^{-1}$ inside the rearing water did not negatively affect the fish, with best growth performance under 40–80 mg $L^{-1}$ $PO_4{}^{3-}$-P. Only highly elevated $PO_4{}^{3-}$-P levels of 120 mg $L^{-1}$ significantly reduce the welfare status. This suggests that limited addition of $PO_4{}^{3-}$-P fertilizer to the plant units in coupled aquaponics with African catfish and the reuse of the plant water in the recirculation aquaculture system are beneficial for both the plants and the fish.

**Author Contributions:** S.M.S.: Experimental planning, main writing, fish performance; A.A.B.: Fish physiology; J.B.: Main sampling, writing; B.B.: Fish welfare; M.O.: Statistical analysis, blood metabolites; B.W.: Product quality; H.W.P.: Experimental planning, writing.

**Funding:** This study was financially supported by the Ministry of Agriculture, Environment and Consumer Protection of Mecklenburg Western Pomerania (BNRZD: 13 903 000 0103) and EIP-AGRI operational groups (WM-EIP-0007−1 5), and the Leibniz Science Campus Phosphorus Research Rostock (SAS-2015-IOW-LWC). We thank the Editors and the reviewers of this paper for their constructive feedback.

**Conflicts of Interest:** The authors declare no conflict of interests.

## References

1.  Becquer, A.; Trap, J.; Irshad, U.; Ali, M.A.; Claude, P. From soil to plant, the journey of P through trophic relationships and ectomycorrhizal association. *Front. Plant Sci.* **2014**, *5*, 548. [CrossRef] [PubMed]

2.  Guillaume, J.; Kaushik, S.; Bergot, P.; Métailler, R. *Nutrition and Feeding of Fish and Crustaceans*; Springer Science & Business Media: Chichester, UK, 2001; ISBN 1-85233-241-7.

3.  Halver, J.E.; Hardy, R.W. (Eds.) *Fish Nutrition*, 3rd ed.; Elsevier: Seattle, WA, USA, 2002; ISBN 978-0-12-319652-1.

4.  Antony Jesu Prabhu, P.; Schrama, J.W.; Kaushik, S.J. Quantifying dietary phosphorus requirement of fish–a meta-analytic approach. *Aquac. Nutr.* **2013**, *19*, 233–249. [CrossRef]

5.  Jones, J.B., Jr. Hydroponics: Its history and use in plant nutrition studies. *J. Plant Nutr.* **1982**, *5*, 1003–1030. [CrossRef]

6.  Resh, H.M. *Hydroponic Food Production: A Definite Guidebook for the Advanced Home Gardener and the Commercial Hydroponic Grower*, 7th ed.; CRC Press: Boca Raton, FL, USA, 2013; ISBN 978-1-43-987867-5.

7.  Palm, H.W.; Knaus, U.; Appelbaum, S.; Goddek, S.; Strauch, S.M.; Vermeulen, T.; Jijakli, M.H.; Kotzen, B. Towards commercial aquaponics: A review of systems, designs, scales and nomenclature. *Aquac. Int.* **2018**, *26*, 318–342. [CrossRef]

8. Somerville, C.; Cohen, M.; Pantanella, E.; Stankus, A. Small scale aquaponic food production. *FAO Fish. Aquac. Tech. Pap.* **2014**, *589*, 11–181.

9. Graber, A.; Junge, R. Aquaponic Systems: Nutrient recycling from fish wastewater by vegetable production. *Desalination* **2009**, *246*, 147–156. [CrossRef]

10. Suhl, J.; Dannehl, D.; Kloas, W.; Baganz, D.; Jobs, S.; Scheibe, G.; Schmidt, U. Advanced aquaponics: Evaluation of intensive tomato production in aquaponics vs. conventional hydroponics. *Agric. Water Manag.* **2016**, *178*, 335–344. [CrossRef]

11. Strauch, S.M.; Wenzel, L.C.; Bischoff, A.; Dellwig, O.; Klein, J.; Schüch, A.; Wasenitz, B.; Palm, H.W. Commercial African catfish (*Clarias gariepinus*) recirculating aquaculture systems: Assessment of element and energy pathways with special focus on the phosphorus cycle. *Sustainability* **2018**, *10*, 1805. [CrossRef]

12. Goddek, S.; Delaide, B.; Mankasingh, U.; Ragnarsdottir, K.V.; Jijakli, H.; Thorarinsdottir, R. Challenges of sustainable and commercial aquaponics. *Sustainability* **2015**, *7*, 4199–4224. [CrossRef]

13. Kloas, W.; Groß, R.; Baganz, D.; Graupner, J.; Monsees, H.; Schmidt, U.; Staaks, G.; Suhl, J.; Tschirner, M.; Wittstock, B.; et al. A new concept for aquaponic systems to improve sustainability, increase productivity, and reduce environmental impacts. *Aquac. Environ. Interact* **2015**, *7*, 179–192. [CrossRef]

14. Razzaque, M.S. Phosphate toxicity: New insights into an old problem. *Clin. Sci.* **2011**, *120*, 91–97. [CrossRef]

15. Ye, C.X.; Liu, Y.J.; Tian, L.X.; Mai, K.S.; Du, Z.Y.; Yang, H.J.; Niu, J. Effect of dietary calcium and phosphorus on growth, feed efficiency, mineral content and body composition of juvenile grouper, *Epinephelus coioides*. *Aquaculture* **2006**, *255*, 263–271. [CrossRef]

16. Oliva-Teles, A.; Pimentel-Rodrigues, A. Phosphorus requirement of European sea bass (*Dicentrarchus labrax* L.) juveniles. *Aquac. Res.* **2004**, *35*, 636–642. [CrossRef]

17. Yang, S.D.; Lin, T.S.; Liu, F.G.; Liou, C.H. Influence of dietary phosphorus levels on growth, metabolic response and body composition of juvenile silver perch (*Bidyanus bidyanus*). *Aquaculture* **2006**, *253*, 592–601. [CrossRef]

18. Ufodike, E.B.C.; Onusiriuka, B.C. Acute toxicity of inorganic fertilizers to African catfish, *Clarias gariepinus* (Teugals). *Aquac. Res.* **1990**, *21*, 181–186. [CrossRef]

19. van Bussel, C.G.J.; Mahlmann, L.; Kroeckel, S.; Schroeder, J.P.; Schulz, C. The effect of high ortho-phosphate water levels on growth, feed intake, nutrient utilization and health status of juvenile turbot (*Psetta maxima*) reared in intensive recirculating aquaculture systems (RAS). *Aquac. Eng.* **2013**, *57*, 63–70. [CrossRef]

20. Palm, H.W.; Knaus, U.; Wasenitz, B.; Bischoff, A.A.; Strauch, S.M. Proportional up scaling of African catfish (*Clarias gariepinus* Burchell, 1822) commercial recirculating aquaculture systems disproportionally affects nutrient dynamics. *Aquaculture* **2018**, *491*, 155–168. [CrossRef]

21. Trejo-Téllez, L.I.; Gómez-Merino, F.C. Nutrient solutions for hydroponic systems. In *Hydroponics-A Standard Methodology for Plant Biological Researches*; Asao, T., Ed.; InTechOpen: London, UK, 2012; pp. 1–22, ISBN 9789535103868.

22. Dorozhkin, S.V. A detailed history of calcium orthophosphates from 1770s till 1950. *Mater. Sci. Eng. C Mater. Biol. Appl.* **2013**, *33*, 3085–3110. [CrossRef] [PubMed]

23. van Rijn, J. Waste treatment in recirculating aquaculture systems. *Aquac. Eng.* **2013**, *53*, 49–56. [CrossRef]

24. Oellermann, L.K. A Comparison of the Aquaculture Potential of *Clarias gariepinus* (Burchell, 1822) and Its Hybrid with *Heterobranchus longifilis* Valenciennes, 1840 in Southern Africa. Ph.D. Thesis, Rhodes University, Grahamstown, South Africa, 1995.

25. Schram, E.; Roques, J.A.C.; Abbink, W.; Spanings, T.; de Vries, P.; Bierman, S.; van de Vis, H.; Flik, G. The impact of elevated water ammonia concentration on physiology, growth and feed intake of African catfish (*Clarias gariepinus*). *Aquaculture* **2010**, *306*, 108–115. [CrossRef]

26. Roques, J.A.; Schram, E.; Spanings, T.; Schaik, T.; Abbink, W.; Boerrigter, J.; de Vries, P.; van de Vis, G.; Flik, G. The impact of elevated water nitrite concentration on physiology, growth and feed intake of African catfish Clarias gariepinus (Burchell 1822). *Aquac. Res.* **2015**, *46*, 1384–1395. [CrossRef]

27. Schram, E.; Roques, J.A.C.; Abbink, W.; Yokohama, Y.; Spanings, T.; Vries, P.; Biermann, S.; van de Vis, H.; Flik, G. The impact of elevated water nitrate concentration on physiology, growth and feed intake of African catfish *Clarias gariepinus* (Burchell 1822). *Aquac. Res.* **2014**, *45*, 1499–1511. [CrossRef]

28. Hogendoorn, H.; Jansen, J.A.J.; Koops, W.J.; Machiels, M.A.M.; van Ewijk, P.H.; van Hees, J.P. Growth and production of the African catfish, *Clarias lazera* (C. & V.): II. Effects of body weight, temperature and feeding level in intensive tank culture. *Aquaculture* **1983**, *34*, 265–285. [CrossRef]

29.   Eding, E.H.; Janssen, K.; Heinsbroek, L.T.N.; Verreth, J.A.J.; Schrama, J.W. Can water phosphorus level in recirculating aquaculture systems (RAS) compensate for low dietary phosphorus level in Nile tilapia Oreochromis niloticus? In Proceedings of the 9th International Conference on Recirculating Aquaculture, Roanoke, VA, USA, 24–26 August 2012.

30.   Hoffman, L.C.; Prinsloo, J.F.; Theron, J.; Casey, N.H. The genotypic influence of four strains of *Clarias gariepinus* on the larvae body proximate, total lipid fatty acid, amino acid and mineral compositions. *Comp. Biochem. Physiol. B Biochem. Mol. Biol.* **1995**, *110*, 589–597. [CrossRef]

31.   Toko, I.I.; Fiogbe, E.D.; Kestemont, P. Mineral status of African catfish (*Clarias gariepinus*) fed diets containing graded levels of soybean or cottonseed meals. *Aquaculture* **2008**, *275*, 298–305. [CrossRef]

32.   Fawole, O.O.; Ogundiran, M.A.; Ayandiran, T.A.; Olagunju, O.F. Proximate and mineral composition in some selected fresh water fishes in Nigeria. *Int. J. Food Saf.* **2007**, *9*, 52–55.

33.   Wedekind, H. Dietary influences on product quality in African catfish (*Clarias gariepinus*). *J. Appl. Ichthyol.* **1995**, *11*, 347–353. [CrossRef]

34.   Rosa, R.; Bandarra, N.M.; Nunes, M.L. Nutritional quality of African catfish *Clarias gariepinus* (Burchell 1822): A positive criterion for the future development of the European production of Siluroidei. *Int. J. Food Sci. Technol.* **2007**, *42*, 342–351. [CrossRef]

35.   Adewumi, A.A.; Adewole, H.A.; Olaleye, V.F. Proximate and elemental composition of the fillets of some fish species in Osinmo Reservoir, Nigeria. *ABJNA* **2014**, *5*, 109–117. [CrossRef]

36.   Wasenitz, B.; Karl, H.; Palm, H.W. Composition and quality attributes of fillets from different catfish species on the German market. *J. Food Saf. Food Qual.* **2018**, *69*, 57–65.

37.   van Dam, A.A.; Pauly, D. Simulation of the effects of oxygen on food consumption and growth of Nile tilapia, *Oreochromis niloticus* (L.). *Aquac. Res.* **1995**, *26*, 427–440. [CrossRef]

38.   Kaushik, S.J.; Doudet, T.; Médale, F.; Aguirre, P.; Blanc, D. Protein and energy needs for maintenance and growth of Nile tilapia (*Oreochromis niloticus*). *J. Appl. Ichthyol.* **1995**, *11*, 290–296. [CrossRef]

39.   Kaushik, S.J.; Médale, F. Energy requirements, utilization and dietary supply to salmonids. *Aquaculture* **1994**, *124*, 81–97. [CrossRef]

40.   Dersjant-Li, Y. Impact of Dietary Cation Anion Difference in Fish and Pigs: A Comparative Study. Ph.D. Thesis, Wageningen University, Wageningen, The Netherlands, 2000.

41.   Al-Kholy, A.; Ishak, M.M.; Youssef, Y.A.; Khalil, S.R. Phosphorus uptake from water by *Tilapia zillii* (Gervais). *Hydrobiologia* **1970**, *36*, 471–478. [CrossRef]

42.   Sugiura, S.H.; McDaniel, N.K.; Ferraris, R.P. In vivo fractional Pi absorption and NaPi-II mRNA expression in rainbow trout are upregulated by dietary P restriction. *Am. J. Physiol. Regul. Integr. Comp. Physiol.* **2003**, *285*, 770–781. [CrossRef] [PubMed]

43.   Sugiura, S.H.; Ferraris, R.P. Dietary phosphorus-responsive genes in the intestine, pyloric ceca, and kidney of rainbow trout. *Am. J. Physiol. Regul. Integr. Comp. Physiol.* **2004**, *287*, 541–550. [CrossRef] [PubMed]

44.   Woo, P.T.K.; Bruno, D.W. (Eds.) *Fish Diseases and Disorders*, 2nd ed.; CABI: Wallingford, UK, 2011; Volume 3, ISBN 978-1-84-593554-2.

45.   Robinson, E.H.; Jackson, L.S.; Li, M.H. Supplemental phosphorus in practical channel catfish diets. *J. World Aquac. Soc.* **1996**, *27*, 303–308. [CrossRef]

46.   Oster, M.; Just, F.; Büsing, K.; Wolf, P.; Polley, C.; Vollmar, B.; Muráni, E.; Ponsuksili, S.; Wimmers, K. Toward improved phosphorus efficiency in monogastrics—Interplay of serum, minerals, bone, and immune system after divergent dietary phosphorus supply in swine. *Am. J. Physiol. Regul. Integr. Comp. Physiol.* **2016**, *310*, 917–925. [CrossRef]

47.   Oster, M.; Gerlinger, C.; Heide, K.; Just, F.; Borgelt, L.; Wolf, P.; Polley, C.; Vollmar, B.; Muráni, E.; Ponsuksili, S.; et al. Lower dietary phosphorus supply in pigs match both animal welfare aspects and resource efficiency. *Ambio* **2018**, *47*, 20–29. [CrossRef]

48.   Ip, Y.K.; Chew, S.F.; Wilson, J.M.; Randall, D.J. Defences against ammonia toxicity in tropical air-breathing fishes exposed to high concentrations of environmental ammonia: A review. *J. Comp. Physiol. B* **2004**, *174*, 565–575. [CrossRef]

49.   Ellis, T.; Yildiz, H.Y.; López-Olmeda, J.; Spedicato, M.T.; Tort, L.; Øverli, Ø.; Martins, C.I. Cortisol and finfish welfare. *Fish Physiol. Biochem.* **2012**, *38*, 163–188. [CrossRef] [PubMed]

50.   Losordo, T.M.; Masser, M.P.; Rakocy, J. Recirculating aquaculture tank production systems. *SRAC* **2000**, *454*, 1–16.

51. Hargreaves, J.A.; Tucker, C.S. Managing ammonia in fish ponds. Stoneville. *SRAC* **2004**, *4603*, 1–8.

52. Rakocy, J.E.; Masser, M.P.; Losordo, T.M. Recirculating aquaculture tank production systems: Aquaponics—Integrating fish and plant culture. *SRAC* **2006**, *454*, 1–16.

53. van de Nieuwegiessen, P.G.; Olwo, J.; Khong, S.; Verreth, J.A.J.; Schrama, J.W. Effects of age and stocking density on the welfare of African catfish, *Clarias gariepinus* Burchell. *Aquaculture* **2009**, *288*, 69–75. [CrossRef]

54. Hontela, A.; Rasmussen, J.B.; Audet, C.; Chevalier, G. Impaired cortisol stress response in fish from environments polluted by PAHs, PCBs, and mercury. *Arch. Environ. Contam. Toxicol.* **1992**, *22*, 278–283. [CrossRef]

55. Fujii, R.; Miyashita, Y. Receptor mechanisms in fish chromatophores–V. MSH disperses melanosomes in both dermal and epidermal melanophores of a catfish (*Parasilurus asotus*). *Comp. Biochem. Physiol. C Comp. Pharmacol. Toxicol.* **1982**, *71*, 1–6. [CrossRef]

56. Sugimoto, M. Morphological color changes in fish: Regulation of pigment cell density and morphology. *Microsc. Res. Tech.* **2002**, *58*, 496–503. [CrossRef]

57. Osborn, C.M. The role of the melanophore-dispersing hormone of the pituitary in the color changes of the catfish. *Proc. Natl. Acad. Sci. USA* **1938**, *24*, 121–125. [CrossRef]

58. Mustapha, M.; Okafor, B.; Olaoti, K.; Oyelakin, O. Effects of three different photoperiods on the growth and body coloration of juvenile African catfish, *Clarias gariepinus* (Burchell). *Arch. Pol. Fish.* **2012**, *20*, 55–59. [CrossRef]

59. Holmberg, K. Ultrastructure and response to background illumination of the melanophores of the Atlantic hagfish, *Myxine glutinosa*, L. *Gen. Comp. Endocrinol.* **1968**, *10*, 421–428. [CrossRef]

60. Dong, X.L.; Lei, W.; Zhu, X.M.; Han, D.; Yang, Y.X.; Xie, S.Q. Effects of dietary oxidized fish oil on growth performance and skin colour of Chinese longsnout catfish (*Leiocassis longirostris* Günther). *Aquac. Nutr.* **2011**, *17*, 861–868. [CrossRef]

61. Agius, C.; Roberts, R.J. Melano-macrophage centres and their role in fish pathology. *J. Fish Dis.* **2003**, *26*, 499–509. [CrossRef]

62. Shao, Q.; Ma, J.; Xu, Z.; Hu, W.; Xu, J.; Xie, S. Dietary phosphorus requirement of juvenile black seabream, *Sparus macrocephalus*. *Aquaculture* **2008**, *277*, 92–100. [CrossRef]

63. Hepher, B. *Nutrition of Pond Fishes*, 1st ed.; Press Syndicate of the University of Cambridge: Cambridge, UK, 1988; ISBN 0521341507.

64. Hosoya, S.; Johnson, S.C.; Iwama, G.K.; Gamperl, A.K.; Afonso, L.O.B. Changes in free and total plasma cortisol levels in juvenile haddock (*Melanogrammus aeglefinus*) exposed to long-term handling stress. *Comp. Biochem. Physiol. A Mol. Integr. Physiol.* **2007**, *146*, 78–86. [CrossRef] [PubMed]

65. Baßmann, B.; Brenner, M.; Palm, H.W. Stress and Welfare of African Catfish (*Clarias gariepinus* Burchell, 1822) in a Coupled Aquaponic System. *Water* **2017**, *9*, 504. [CrossRef]

66. Hecht, T.; Uys, W.; Britz, P.J. The culture of Sharptooth Catfish, *Clarias gariepinus* in Southern Africa. *S. Afr. Natl. Sci. Program. Rep.* **1988**, *153*, 47–61.

67. Belão, T.C.; Leite, C.A.C.; Florindo, L.H.; Kalinin, A.L.; Rantin, F.T. Cardiorespiratory responses to hypoxia in the African catfish, *Clarias gariepinus* (Burchell 1822), an air-breathing fish. *J. Comp. Physiol. B* **2011**, *181*, 905. [CrossRef]

68. Britz, P.J.; Pienaar, A.G. Laboratory experiments on the effect of light and cover on the behaviour and growth of African catfish, *Clarias gariepinus* (Pisces: Clariidae). *J. Zool.* **1992**, *227*, 43–62. [CrossRef]

69. Zayed, A.E.; Mohamed, S.A. Morphological study on the gills of two species of fresh water fishes: *Oreochromis niloticus* and *Clarias gariepinus*. *Ann. Anat.* **2004**, *186*, 295–304. [CrossRef]

*fishes*

MDPI

*Article*

# Influence of Age on Stress Responses of White Seabream to Amyloodiniosis

**Márcio Moreira** [1,2], **Anaísa Cordeiro-Silva** [1], **Marisa Barata** [1], **Pedro Pousão-Ferreira** [1] and **Florbela Soares** [1,*]

1    IPMA-Portuguese Institute for the Ocean and Atmosphere, EPPO-Aquaculture Research Station,
     8700-305 Olhão, Portugal; marcio.moreira@ipma.pt (M.M.); anaisa.silva@ipma.pt (A.C.-S.);
     mbarata@ipma.pt (M.B.); pedro.pousao@ipma.pt (P.P.-F.)
2    CCMAR, Centre of Marine Sciences, University of Algarve, 8005-139 Faro, Portugal
*    Correspondence: fsoares@ipma.pt

Received: 24 February 2019; Accepted: 4 April 2019; Published: 8 April 2019

check for updates

**Abstract:** Amyloodiniosis is a disease that represents a major bottleneck for semi-intensive aquaculture, especially in Southern Europe. The inefficacy of many of the treatments for this disease on marine fish produced in semi-intensive aquaculture has led to a new welfare approach to amyloodiniosis. There is already some knowledge of several welfare issues that lead to amyloodiniosis as well as the stress, physiological, and immunological responses to the parasite by the host, but no work is available about the influence of fish age on the progression of amyloodiniosis. The objective of this work was to determine if stress, hematological, and histopathological responses are age dependent. For that purpose, we determined the mortality rate, histopathological lesions, hematological indexes, and stress responses (cortisol, glucose, lactate, and total protein) in "Small" (total weight: $50 \pm 5.1$ g, age: 273 days after eclosion (DAE)) and "Big" (total weight: $101.3 \pm 10.4$ g, age: 571 DAE) white seabream (*Diplodus sargus*) subjected to an *Amyloodinium ocellatum* infestation (8000 dinospores $mL^{-1}$) during a 24-h period. The results demonstrated a strong stress response to *A. ocellatum*, with marked differences in histopathological alterations, glucose levels, and some hematological indexes between the fish of the two treatments. This work elucidates the need to take in account the size and age of the fish in the development and establishment of adequate mitigating measures and treatment protocols for amyloodiniosis.

**Keywords:** aquaculture; *Amyloodinium ocellatum*; age; physiological response; hematology; histopathology; welfare

## 1. Introduction

Amyloodiniosis is a disease that represents a major bottleneck for semi-intensive aquaculture, especially in Southern Europe [1]. It is caused by one of the most common and important parasitic dinoflagellates in marine fish, *Amyloodinium ocellatum* (Brown). This parasite can potentially affect almost all fish species living within its ecological range (temperature: 16–30 °C; salinity: 10–60 psu) [2]. *A. ocellatum* has a triphasic lifecycle with a parasitic trophont, a reproductive encysted tomont, and a free-motile infective dinospore stage [3]. It is considered one of the most consequential pathogens for marine fish, causing serious morbidity and mortality (around 100%) in brackish and marine warmwater fish in different aquaculture facilities worldwide [2]. The affected producers can suffer severe economic impacts, as demonstrated in reported cases of amyloodiniosis in a milkfish (*Chanos chanos*) hatchery in 2004 [4] or in Nile tilapia (*Oreochromis mossambicus*) in the Salton Sea [5], where total losses reached US$20,000 and US$6–77 million, respectively [6]. The open design of many aquaculture systems also allows the easy dissemination of this parasite to new places in their ecological range, where they find ideal conditions to cause disease outbreaks [7,8]. *A. ocellatum* outbreaks develop extremely fast and, at the time of its detection, contaminated fish no longer respond to treatment, resulting in 100% mortality in a few days [1]. The clinical signs of this disease are changes in fish behavior, with jerky movements, swimming at the water surface, and decreased appetite [1]. These may also include increased respiratory rate and gathering at the surface or in areas with higher dissolved oxygen concentrations.

There are several treatments available for amyloodiniosis, especially for the motile infective dinospore [9,10], since the parasitic and encapsulated state of trophont and tomont are difficult to treat and eradicate [1]. These treatments can range from the most classical ones, such as formaldehyde [11]; copper sulphate [1,10,12,13], which can be done in combination with freshwater treatment [14]; or hydrogen peroxide [15,16], to the more atypical ones, such as the addition of larval brine shrimp (*Artemia salina*) to the tank to prey on *A. ocellatum* dinospores [17]. However, most of these treatments are highly ineffective or unpractical for earthen pond semi-intensive aquaculture, which is the most common type of aquaculture facility in Southern Europe [1]. This treatment problem, considering that this is a common and highly consequential disease for producers and that naturally occurring *A. ocellatum* blooms have already been detected [18,19], led us to make a new and more holistic approach to this disease.

One of the more promising approaches to amyloodiniosis is to consider that it is not only a health and epidemiological problem, which is a very reductionistic approach to the disease, but also a welfare problem [20].

A disease as a welfare problem is a very special case because it is caused by several other welfare issues, such as inadequate aquaculture protocols [21], poor water quality [22], nutritional imbalances [23], physical disturbances (e.g., transport [24] or handling [25]), stocking densities [26], or social behaviors [27] that can increase the probability of disease outbreaks. However, the disease is, by itself, a welfare problem and can also cause other welfare problems, such as a decrease in water quality due to fish death and decay as well as secondary effects of fish treatment [28]. All of these pre- and post-disease welfare issues can cause chronic stress [24] and immunocompetence loss in fish [29].

Regarding amyloodiniosis, some work has already been done on controlling some of the welfare issues that can cause this disease. We already know the temperature, oxygen, and salinity conditions that favor amyloodiniosis outbreak [30–32]. Previous works have already established the water quality parameters, fish stock density, and water renewal rate in fish production ponds to avoid *A. ocellatum* infestation in seabream [33]; the consequences for water quality due to fish death and decay [16]; and routine hygiene procedures to avoid water quality problems and reinfestation during an amyloodiniosis outbreak [1]. There are also works that state that fish mortality in an amyloodiniosis outbreak can be caused by anoxia, which is associated with serious gill hyperplasia, inflammation, hemorrhage, and necrosis, in heavy infestations [9] or osmoregulatory impairment and secondary microbial infections due to severe epithelial damage in some lethal cases, which is associated with apparently mild infestations [31]. Further, the immunological [34–39], stress, and metabolic responses of some fish species to *A. ocellatum* [40–44] have also been studied.

However, there have been no studies on the effect of age on the responses or mortality of fishes to an *A. ocellatum* infestation.

So, the objective of this work was to analyze the stress and hematological responses and mortality of white seabream (*Diplodus sargus*) from different ages when exposed to an *A. ocellatum* infestation.

## 2. Results

### 2.1. Gill Analysis

White seabream gill wet mounts from Small and Big treatments are presented in Figure 1.

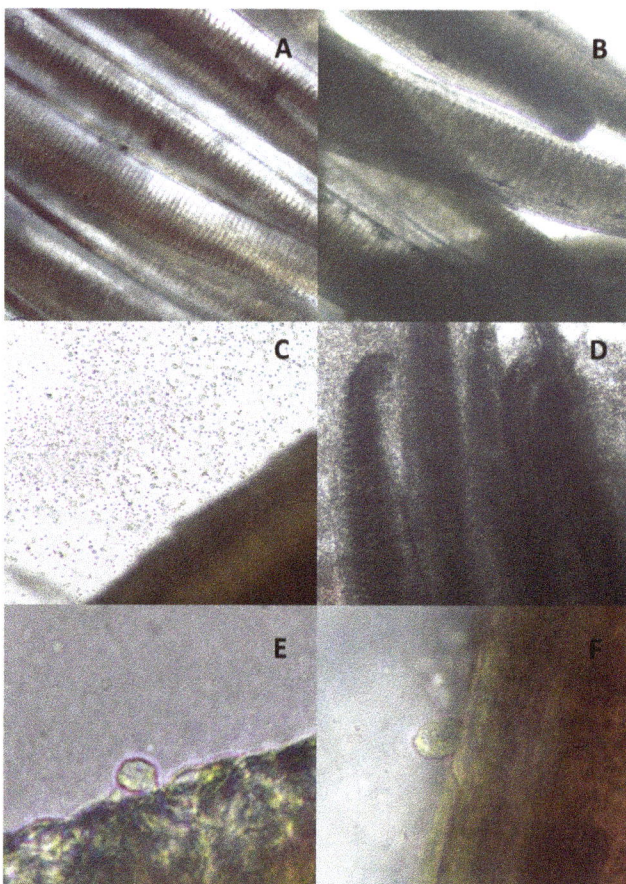

**Figure 1.** Wet mount of white seabream (*Diplodus sargus*) gills from "Small" and "Big" experimental groups: (**A**) gill from Small treatment at 0 h (100×); (**B**) gill from Big treatment at 0 h (100×); (**C**) Small treatment at 5 h, presenting several *Amyloodinium ocellatum* dinospores near the gill (400×); (**D**) Big treatment at 24 h, presenting several *A. ocellatum* dinospores near the gill and a great amount of mucus (400×); and *A. ocellatum* trophonts in gill from Small treatment (**E**) and Big treatment (**F**) at 5 h of infestation.

We observed that there were no parasites in the gills of both treatments at 0 h of white seabream *A. ocellatum* infestation, which validated the history of no parasitological infestation of the fishes (see Figure 1A,B).

Five hours after infestation, we observed the fixation of *A. ocellatum* dinospores and their transformation to trophonts in both treatments (see Figure 1C,E,F). However, the *A. ocellatum* trophont counts revealed that the Small treatment had more parasite fixation (more than 500 parasites per gill arch) than the Big treatment (between 250 and 300 parasites per gill arch).

Twenty-four hours after infestation, we also observed the secretion of massive quantities of mucus by the gill in the Big treatment (see Figure 1D). The parasite count was $155 \pm 5$ parasites for the gill arch, which was lower than the parasite load observed after 5 h of infestation for this treatment. The gills from the Small treatment at 24 h were too degraded for a viable parasite count.

## 2.2. Mortality

After 5 h of infestation by the parasite *A. ocellatum*, both treatments had a white seabream mortality of 8.3%, which reached 100% in the Small treatment and 33.3% in the Big treatment after 24 h of infestation.

## 2.3. Histopathological Analysis

In the histological analysis of gills, several differences were observed between Small and Big treatments when white seabream were infested with *A. ocellatum*. The types of lesions observed in the gills of white seabream are represented in Figure 2.

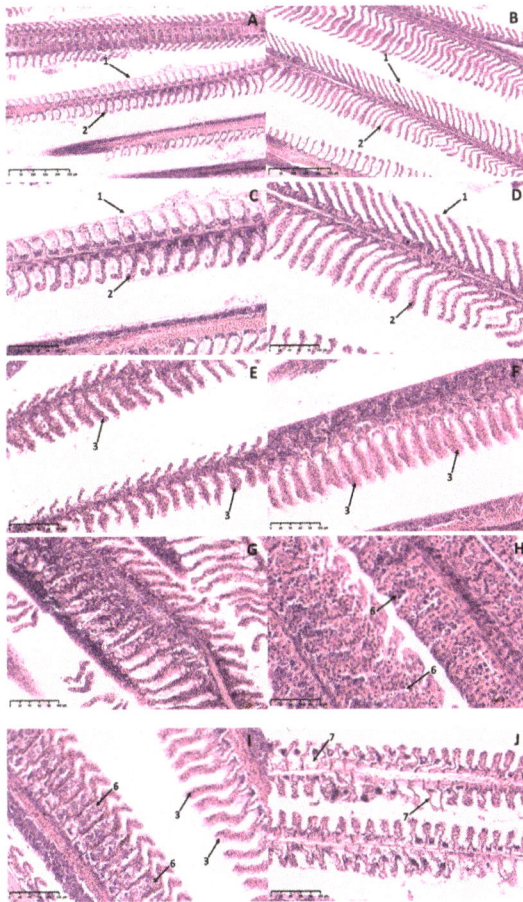

**Figure 2.** Histological section of hematoxylin–eosin (H&E) stained gills from white seabream (*D. sargus*) during an *A. ocellatum* infestation. (**A**) (100×) and (**B**) (200×) represent the Small and (**C**) (100×) and (**D**) (200×) represent the Big treatments at 0 h, with normal gills (1—primary lamella; 2—secondary lamella); (**E**) (200×) and (**G**) (200×) represent the Small and (**F**) (200×) and (**H**) (400×) represent the Big treatments at 5 h of infestation. *A.-ocellatum*-infested fish gills with several histopathological alterations: hyperplasia of the lamellar epithelium (3), vacuolization (4), mucus cells (5), and fusion of secondary lamellae (6); (**I**) (200×) and (**J**) (200×) images represent *A.-ocellatum*-infested fish gills of the Big treatment after 24 h of infestation, with several histopathological alterations: hyperplasia of the lamellar epithelium (3), fusion of secondary lamellae (6), and necrosis (7).

The images of Figure 2 show that *A. ocellatum* infestation led to histopathological alterations such as hyperplasia and vacuolization of the lamellar epithelium and secondary lamellae fusion.

We observed that there were differences in the type of histopathological alterations between treatments 5 h after treatment. Gills of the Small treatment presented hyperplasia and vacuolization of the lamellar epithelium. However, gills of the Big treatment presented hyperplasia of the lamellar epithelium, an increase of mucus cells, and fusion of secondary lamellae, which culminated in gill necrosis for this treatment at 24 h of *A. ocellatum* infestation. The gills from the Small treatment at 24 h were too degraded for histopathological analysis.

### 2.4. Stress Indicators

The variations of the cortisol, glucose, lactate, and total protein levels in white seabream exposed to an *A. ocellatum* outbreak are presented in Figure 3.

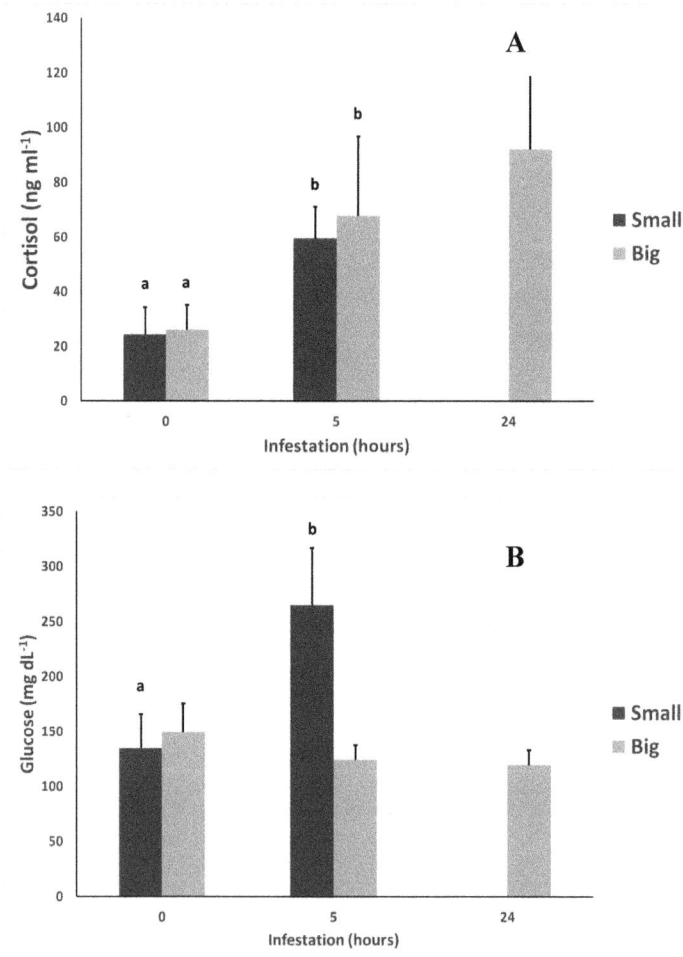

**Figure 3.** *Cont.*

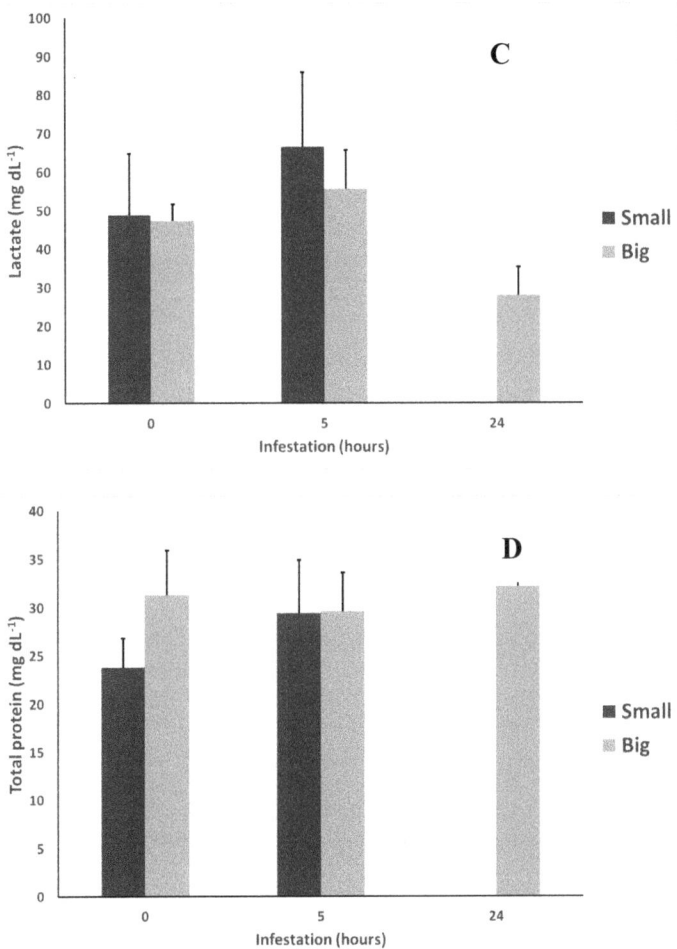

**Figure 3.** White seabream (*D. sargus*) cortisol (**A**), glucose (**B**), lactate (**C**), and total protein (**D**) levels in Small and Big treatments during an *A. ocellatum* infestation (N = 12 at 0 h after infestation for both treatments, N = 11 at 5 h after infestation for both treatments, N = 0 for Small treatment, and N = 8 for Big treatment at 24 h after infestation. Error bars = standard deviation). Significant statistical differences between treatments ($p < 0.05$) correspond to a and b letters.

We can observe that there were significant differences ($p < 0.05$) in the cortisol levels between fish from 0 to 5 h after infestation in both treatments. The Big treatment also presented significant differences ($p < 0.05$) in glucose levels between 0 and 5 h after infestation.

Lactate and total protein did not present any statistically significant differences ($p < 0.05$) between treatments or along the time of the experiment.

*2.5. Hematological Analysis*

The results of the different hematological indicators analyzed are shown in Table 1.

Table 1. Hematological indicators in Small and Big treatments during an *A. ocellatum* infestation on white seabream (*D. sargus*) at 0, 5, and 24 h after infestation (N = 12 at 0 h after infestation for both treatments, N = 11 at 5 h after infestation for both treatments, N = 0 for Small treatment, and N = 8 for Big treatment at 24 h after infestation. Error bars = standard deviation). Significant statistical differences between treatments ($p < 0.05$) correspond to a and b letters.

| Hematological Indicators | Small | | | Big | | |
|---|---|---|---|---|---|---|
| | 0 | 5 | 24 | 0 | 5 | 24 |
| Hct (% ± SD) | 29.7 ± 5.90 | 34.5 ± 6.47 | NA | 31.2 ± 3.74 | 32.5 ± 3.80 | 30.4 ± 4.80 |
| Hemoglobin (g dL$^{-1}$ ± SD) | 5.24 ± 0.85 | 5.38 ± 1.21 | NA | 6.39 ± 0.74 | 6.93 ± 0.82 | 6.42 ± 0.78 |
| RBC (% ± SD) | 99.9 ± 1.38 | 99.9 ± 0.54 | NA | 99.3 ± 1.30 | 99.6 ± 1.57 | 99.6 ± 1.52 |
| WBC (% ± SD) | 0.081 ± 0.030 [a] | 0.034 ± 0.014 [b] | NA | 0.703 ± 0.151 [a] | 0.422 ± 0.082 [b] | 0.430 ± 0.141 [b] |
| MCV (fl ± SD) | 73.6 ± 1.46 | 75.1 ± 1.41 | NA | 71.4 ± 1.47 [a] | 80.0 ± 1.87 [b] | 82.5 ± 1.79 [b] |
| MCH (pg ± SD) | 13.0 ± 2.11 | 11.7 ± 2.64 | NA | 14.6 ± 1.48 | 17.0 ± 2.01 | 17.3 ± 2.11 |
| MCHC (gHb 100mL$^{-1}$ ± SD) | 1.76 ± 0.29 | 1.56 ± 0.35 | NA | 2.05 ± 0.21 | 2.13 ± 0.25 | 2.10 ± 0.26 |

*Hct—Hematocrit; RBC—Red Blood Cell count; WBC—White Blood Cell count; MCV—Mean Corpuscular Volume; MCH—Mean Corpuscular Hemoglobin; MCHC—Mean Corpuscular Hemoglobin Concentration; NA—sample not available due to fish death.

Hematocrit (Hct), hemoglobin, and red blood cells (RBC) did not present any statistical differences ($p < 0.05$) between Small and Big white seabream during the experiment, even if there was a tendency for an increase of this values over time. Mean corpuscular hemoglobin (MCH) and mean corpuscular hemoglobin concentration (MCHC) also did not present any statistical differences ($p < 0.05$) between treatments.

However, we observed statistical differences ($p < 0.05$) in white blood cell (WBC) values between treatments, with a lower percentage of WBC in the Small treatment. There were also statistical differences ($p < 0.05$) between 0 and 5 h after infestation in both treatments, with a decrease in the percentage of WBC over time.

For mean corpuscular volume (MCV) values, there were statistical differences ($p < 0.05$) in the Big treatment between 0 and 5 h after infestation, with an increase in MCV values over the time.

## 3. Discussion

Amyloodiniosis causes mortality mainly due to anoxia, osmoregulatory impairment, and secondary microbial infections due to severe epithelial damage [9,31]. This can trigger several physiological responses, as seen for gilthead seabream (*Sparus aurata*) [41], European sea bass (*Dicentrarchus labrax*) [39], and yellowtail (*Seriola dorsalis*) [43]. However, for white seabream, there are no references about physiological responses to *A. ocellatum* infestations.

The gill wet mount observation revealed that the gills in both treatments (Small and Big) did not have any parasite colonization at the beginning of the experiment, which confirms the absence of parasitological contamination in the used white seabream. It also demonstrated that at 5 and 24 h after infestation, there was *A. ocellatum* dinospore fixation and trophont development in both treatments.

The histopathological analysis of the white seabream gills revealed an increase in some histopathological lesions, such as hyperplasia with vacuolization, lamellar fusion, and necrosis. This agrees with the lesions reported for several fish species such as cobia (*Rachycentron canadum*) [40], porkfish (*Anisotremus virginicus*) [45], gilthead seabream [42], yellowtail [43], Senegalese sole (*Solea senegalensis*) [46], meagre (*Argyrosomus regius*) [47], silver pompano (*Trachinotus blochii*) [48], and European sea bass [38] for infestation of *A. ocellatum*. However, there were some differences in the type of histopathological lesions over time. Small treatment gills presented hyperplasia and vacuolization of the lamellar epithelium, and Big treatment gills presented hyperplasia of the lamellar epithelium, an increase in mucus cells (which was confirmed by the observation of great quantities of mucus in the gills for this treatment on the gill wet mount observations), and fusion of secondary lamellae at 5 h after infestation. The vacuolization of gill tissue observed at 5 h of infestation in the Small treatment is usually associated with detachment of the respiratory epithelium and necrosis [31], which can explain the 100% mortality observed at 24 h after infestation for this treatment. The increase in mucus cells observed at 5 h of infestation for the Big treatment corresponded to a gill mucosa-associated lymphoid tissue (MALT) mechanism of defense used by fish as a general response to ectoparasites [49–51], which can also lead to gill dysfunction due to excess mucus production [43]. This could explain the gill necrosis observed at 24 h of infestation for the Big treatment as well as the lower mortality rate in comparison with the Small treatment. The absence of an increase of mucus cells observed at 5 h of infestation for the Small treatment may be due to a certain immaturity of the MALT system in younger fish [49].

As natural stressors, parasites can trigger several stress responses in fish. One of the most used indicators for parasitologically induced stress response is cortisol [52]. In this experiment, a statistically significant ($p < 0.05$) higher concentration of cortisol in the fish was observed for the two treatments at 5 h of infestation and was maintained at 24 h of infestation in the Big treatment. This agrees with the cortisol levels reported in gilthead seabream and sea bass exposed to *A. ocellatum* [39,41] and is within the range of cortisol values reported for stressed white seabream [53].

As a consequence, cortisol can induce higher concentrations of glucose caused by extended glycogenolysis and gluconeogenesis through the degradation of glycogen [54]. In this experiment, we did find statistical differences in glucose levels between 0 and 5 h of infestation for the Big treatment but not for the Small treatment. The glucose values were within the range of glucose values reported for white seabream [55,56]. The statistical difference observed in glucose levels in the Big treatment at 5 h after infestation are in agreement with the data observed for yellowtail infested with *A. ocellatum* [43] and could be a consequence of lipid storage usage for energy expenditure on gill mucus production and osmoregulation [39]. However, the absence of statistical differences in glucose levels between 0 and 5 h of infestation in the Small treatment is puzzling, even if it is in agreement with glucose levels observed in European seabass infested with *A. ocellatum*, which suggests a possible inhibition of the glycogenolysis and gluconeogenesis pathways in the liver by *A. ocellatum* [39]. This disparity of response between the two treatments could also be related to fish size [39].

The lactate values observed in this experiment were within the range of values reported for white seabream [55]. The absence of statistical differences in both treatments for lactate levels is not in agreement with the differences observed in other studies with fish parasites [57]. However, there was a tendency for a high level of lactate at 5 h after infestation. This could be due to the role of lactate as a response to anoxia due to severe gill epithelial damage by *A. ocellatum* [9,31] acting in hypoxia signaling and collagen deposition processes. This also explains the tendency of higher levels of lactate in the Small treatment at 5 h of infestation, proportional to the gill lesions observed in the histopathological analysis for this treatment.

There were no statistical differences in both treatments for total protein levels. This is in agreement with the absence of statistical differences in total protein levels observed for gilthead seabream [42]. This could be due to the consequences of high cortisol levels, which control the blood osmolality and pH by regulating the balance of blood potassium and sodium ions coming from the higher production of other metabolites and proteins in response to stress [58].

In this experiment, hematological responses showed that Hct values were within the range of values expected for white seabream [59] and they were not statistically different between treatments. This agrees with the data observed for gilthead seabream in an infestation with *A. ocellatum* [41].

The results obtained for hemoglobin and RBC showed no statistical differences between treatments. This does not agree with the hemoglobin and RBC data observed in yellowtail infested with *A. ocellatum* [43]. However, the results obtained for the hemoglobin and RBC analysis followed a pattern similar to Hct, which is in agreement with Horton and Okamura's [60] observations for infested fishes.

The WBC profile can undergo significant changes during fish disease outbreaks and is thus an excellent indicator of immune responses. A significant decrease in WBC was observed in both treatments at 5 h of infestation, indicating a possible inhibition of the immune system due to high levels of cortisol [61]. This is in agreement with the data observed for gilthead seabream infested with *A. ocellatum* [41] and could be also associated with a possible mechanism of innate immune system evasion by this parasite [50]. Further studies are necessary to better evaluate the response of white seabream to an *A. ocellatum* infestation.

MCV showed an increasing tendency but without any statistical differences in the Small treatment, but there were statistical differences in the Big treatment between 0 and 5 h after infestation, with an increase in MCV values over time. This is in agreement with data obtained for this index in parasitic infections [62]. The absence of MCH and MCHC is also in agreement with the previous data.

When we look at all the obtained results, we can observe that the age of white seabream seems to affect the host response to an *A. ocellatum* infestation. One possible interpretation of the results is presented on Figure 4.

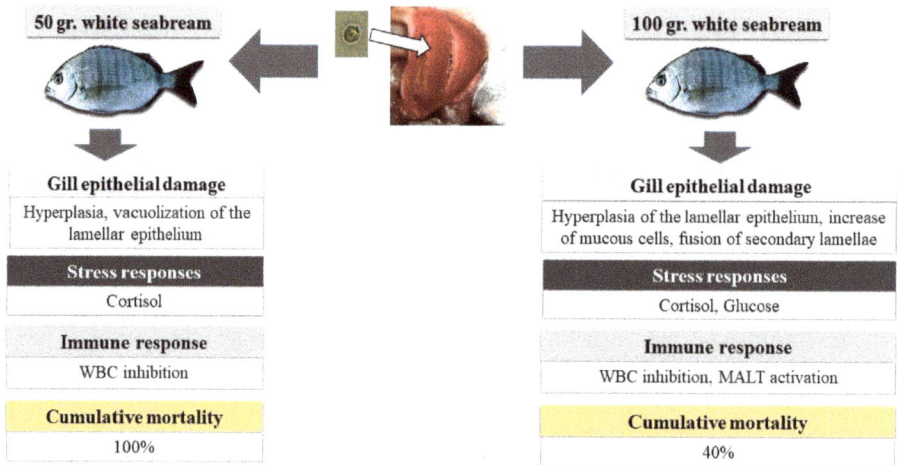

| 50 gr. white seabream | | 100 gr. white seabream |
|---|---|---|

| **Gill epithelial damage** | | **Gill epithelial damage** |
|---|---|---|
| Hyperplasia, vacuolization of the lamellar epithelium | | Hyperplasia of the lamellar epithelium, increase of mucous cells, fusion of secondary lamellae |
| **Stress responses** | | **Stress responses** |
| Cortisol | | Cortisol, Glucose |
| **Immune response** | | **Immune response** |
| WBC inhibition | | WBC inhibition, MALT activation |
| **Cumulative mortality** | | **Cumulative mortality** |
| 100% | | 40% |

**Figure 4.** Possible interpretation of the results obtained from the stress, hematological, and histopathological analysis of Small (50 ± 5.1 g, age: 273 days after eclosion (DAE)) and Big (101.3 ± 10.4 g, age: 571 DAE) white seabream (*D. sargus*) during an *A. ocellatum* infestation.

So, when *A. ocellatum* dinospores reached the gill and started the fixation and transformation into trophonts, both Small and Big white seabream began to present lesions in gill tissue which increased over time. However, these were more severe in the Small treatment, possibly due to the nonactivation of the MALT defense system, which was present in the Big treatment (increase in gill mucus cells). This triggered a primary stress response in both treatments (cortisol), with a secondary stress response (glucose) in the Big treatment. However, the absence of glucose and lactate response in the Small treatment indicates a possible inhibition of the glycogenolysis and gluconeogenesis pathways in the liver in the Small treatment. The primary stress response to *A. ocellatum* also inhibited the WBC immune response in both treatments. These differences in the response to an *A. ocellatum* outbreak by the Small and Big treatments may explain the differences in mortality 24 h after infestation.

The findings of this paper highlight the necessity of considering fish age when designing vigilance plans (which are designed nowadays mainly taking into account the water temperature) and determining treatment actions during an *A. ocellatum* infestation. In this way, younger fish will need to have a stricter vigilance plan, with short ectoparasitological sampling periods and mitigation measures and treatment application at the first *A. ocellatum* trophont detection, to reduce fish stress and mortality resulting from amyloodiniosis. Older fish could continue to have a less strict vigilance plan, with less frequent ectoparasitological sampling periods and application of mitigation measures at lower infestations of *A. ocellatum*, and treatment application only after *A. ocellatum* trophont numbers reach a certain threshold, which should be determined by taking into account the water temperature of the tank.

## 4. Materials and Methods

### 4.1. Fish Culture Conditions

White seabream juveniles were reared at EPPO-IPMA (Aquaculture Research Centre, National Institute for the Sea and Atmosphere, Olhão, Portugal) and originated from a wild broodstock. After hatching, fish were reared following the protocol used in EPPO-IPMA for this species [63].

Two sets of 36 individuals, one with an average total weight of $50 \pm 5.1$ g and an age of 273 days after eclosion (DAE) and the other with an average total weight $101.3 \pm 10.4$ g and an age of 571 DAE, with no history of ectoparasites and no incidence of malformations, were selected. The fish were kept in two separate $1.5$-m$^3$ tanks according to their weight (36 per tank) with a 14 h light/10 h dark photoperiod (photophase from 6:00 a.m. to 8:00 p.m.). Water temperature was $20 \pm 1\,^{\circ}C$ and water salinity was 37.5 psu. The fish were full-fed with commercial feed three times a day until satiation.

### 4.2. Preparation of A. Ocellatum Infection Tanks

Eighteen 80-L rectangular plastic tanks were infested with 1000–1200 *A. ocellatum* tomonts, obtained from an induced outbreak according to the methodology described in [42], and then incubated until reaching an average of 8000 dinospores/mL to maximize the stress response [41]. The water temperature was maintained at $24 \pm 0.2\,^{\circ}C$, 36.5 psu of salinity, and 100% dissolved oxygen in closed seawater systems, artificial aeration, and a 14 h light/ 10 h dark photoperiod. To avoid differences between tanks, the water from the tanks were mixed before fish were introduced to ensure the parasite load was equal in every tank.

### 4.3. Experimental Design

Fish were fasted for 24 h prior to the experiment. Four fish (N = 4) were measured, weighed, and then transferred to each 80 L tank (Small treatment—white seabream weighing $50 \pm 5.1$ g, and Big treatment—white seabream weighing $101.3 \pm 10.4$ g). Each treatment had tanks in triplicate for all the sampling points. Fish were sampled at 5 and 24 h after the beginning of the experiment.

Mortality at each point of sampling was also registered, and dead fish were withdrawn from the tanks before fish sampling.

### 4.4. Sampling Protocol

Twelve fish per treatment were sampled in the beginning (0 h) of the experiment. At 5 and 24 h of the experiment, all the surviving fish in each treatment were sampled, prioritizing the most moribund fish, according to their behavior [1]. Fish that presented erratic and slower swimming movements, fish aggregation near the aerators and water line, accelerated opercular and mouth opening movements, or reddish marks on the skin were sampled first.

After sedation with 800 ppm of 2-phenoxyethanol [64], blood was withdrawn from the caudal vein with a 1 mL heparinized syringe. Collection of blood samples (1 mL of blood per fish, on average) was completed within 3 min after capturing the fish to minimize the effects of handling on stress parameters. The collected blood from each fish was divided in two aliquots. One of the aliquots was immediately frozen at $-80\,^{\circ}C$ for posterior analyses. The other blood aliquot was used for hematological analysis.

After blood collection, the fish was sacrificed with an overdose of 2-phenoxyethanol, according with the EU Directive 2010/63/EU for animal experimentation.

Plasma was separated from cells by centrifugation (10 min, 5000× $g$, room temperature), snap-frozen in liquid nitrogen, and stored at −80 °C until analysis of osmolarity, pH, hormonal (cortisol), and metabolic (glucose, lactate, and protein concentrations) parameters.

Then, the parasite burden was evaluated by microscopic observation of a wet mount of the first two branchial arches from fish killed by anesthetic overdose with 2-phenoxyethanol [41] from each treatment at all sampling points.

### 4.5. Histopathological Analysis

For histological analysis, we collected the first gill from all the fishes from each tank (N = 12 per treatment). Following fixation in Davidson's solution (pH 7.2) for two days, tissues were transferred to 70% ethanol. The tissues were processed in an automatic tissue processor Leica TP 1020 and included in paraffin wax block. Sections that were 5-nm thick were cut with a microtome slide Model Leica 5M 2000 R and stained with hematoxylin–eosin (H&E) according to the procedure described in [65]. Gill tissue was examined under a Nikon H550S microscope using bright-field illumination for the presence of *A. ocellatum* and possible pathological changes. Selected gill tissues were then scanned in a Hamamatsu Nano Zoomer Digital Pathology, and images were taken and processed using NDP View 2 software.

### 4.6. Stress Indicator Analysis

Plasma glucose and lactate levels were measured using commercial kits from Spinreact (Glucose-HK Ref. 1001200; Lactate Ref. 1001330) adapted to 96-well microplates. Plasma proteins were determined in diluted plasma samples 1:5 (v/v) with the QCA Total Proteins kit (Química Clínica Aplicada S.A., Barcelona, Spain). Plasma cortisol levels were quantified by an ELISA kit (Cortisol ELISA kit, IBL) modified and adapted to fish [66]. Cortisol was extracted from 20 µL of plasma in 200 µL of diethyl ether. The lower limit of detection (81% binding) was 0.1 ng mL$^{-1}$ plasma. All the samples were examined in duplicate and analyzed on a microplate reader (Multiskan GO, Thermo Scientific).

### 4.7. Hematological Analysis

Hct determination was performed as described by Soares et al. [67]. In summary, a blood sample portion was transferred to nonheparinized microhematocrit tubes and centrifuged at 5000× $g$ for 5 min with a hematocrit centrifuge (EBA 21 Hettich) to determine the hematocrit value. The determination of hematocrit was calculated as the percentage of red blood cells present in the amount of blood collected.

Total RBC and WBC were counted from blood dilutions using a hemocytometer.

Hemoglobin (Hgb) determination was performed using a commercial kit (1001230, Spinreact) based on the Drabkin colorimetric principle. All samples were examined in duplicate and analyzed on a microplate reader at 540 nm (Multiskan GO, Thermo Scientific) [68].

MCV, MCH, and MCHC were calculated with the formula described by Radu et al. [69].

### 4.8. Statistical Analysis

For group comparisons, normality was assessed using the Shapiro–Wilk test, while the homogeneity of variance was assessed using Bartlett's test. If the data were parametric, nested ANOVA followed by Tukey's test was done to detect differences between Small and Big fish. Statistical significance was accepted at $p < 0.05$. These statistical tests were made using R studio (version 1.0.153). Values are expressed as mean ± SD.

*Fishes* **2019**, *4*, 26

## 5. Conclusions

We can conclude that the age of white seabream affects the response to an *A. ocellatum* infestation, especially at stress and histopathological levels, since smaller and younger fish seemed to have a more immature MALT defense system, and the stress reaction to the parasite seemed to be more acute. This study demonstrates that age is an influential factor in the response to an *A. ocellatum* infestation in white seabream, especially at stress and histopathological levels, and elucidates the need to take into account the size and age of the fish in the development and establishment of adequate mitigating measures and treatment protocols for amyloodiniosis. We recommend the establishment of a stricter vigilance protocol with scheduled ectoparasitological samples for younger fish, since they are more susceptible to an *A. ocellatum* outbreak. This will allow for detection of the parasitosis at an early stage of development and quicker treatment, which will avoid high mortality levels and high levels of fish stress.

**Author Contributions:** M.M.: Conceptualization, Methodology, Formal Analysis, Investigation, Writing—Original Draft, Visualization. A.C.-S.: Resources, Writing—Review and Editing. M.B.: Resources, Writing—Review and Editing. P.P.-F.: Project Administration, Funding Acquisition. F.S.: Conceptualization, Validation, Writing—Review and Editing, Supervision.

**Funding:** This study was funded by Projects DIVERSIAQUA (Mar2020 16-02-01-FMP-0066) and SAUDE&AQUA (Mar2020 02.05.01-FEAMP-0009). This study also received Portuguese national funds from FCT—Foundation for Science and Technology through project UID/Multi/04326/2019.

**Acknowledgments:** Márcio Moreira has a PhD grant from FCT (SFRH/BD/118601/2016). We would like to thank Janine Sieben for her help in editing the English language of the manuscript.

**Conflicts of Interest:** The authors declare no conflict of interests and the funders had no role in the design of the study; in the collection, analyses, or interpretation of data; in the writing of the manuscript; or in the decision to publish the results.

## References

1. Soares, F.; Quental Ferreira, H.; Cunha, E.; Pousão-Ferreira, P. Occurrence of *Amyloodinium ocellatum* in aquaculture fish production: A serious problem in semi-intensive earthen ponds. *Aquac. Eur.* **2011**, *36*, 13–16.
2. Paperna, I.; Ross, B.; Colorni, A.; Colorni, B. *Diseases of Marine Fish Cultured in Eilat Mariculture Project Based at the Gulf of Aqaba, Red Sea*; 92-5-000964-X; FAO: Rome, Italy, 1980; pp. 29–32.
3. Landsberg, J.H.; Steidinger, K.A.; Blakesley, B.A.; Zondervan, R.L. Scanning Electron-Microscope Study of Dinospores of *Amyloodinium* cf *Ocellatum*, a Pathogenic Dinoflagellate Parasite of Marine Fish, and Comments on Its Relationship to the Peridiniales. *Dis. Aquat. Org.* **1994**, *20*, 23–32. [CrossRef]
4. Cruz-Lacierda, E.R.; Maeno, Y.; Pineda, A.J.T.; Matey, V.E. Mass mortality of hatchery-reared milkfish (*Chanos chanos*) and mangrove red snapper (*Lutjanus argentimaculatus*) caused by *Amyloodinium ocellatum* (Dinoflagellida). *Aquaculture* **2004**, *236*, 85–94. [CrossRef]
5. Kuperman, B.I.; Matey, V.E. Massive infestation by *Amyloodinium ocellatum* (Dinoflagellida) of fish in a highly saline lake, Salton Sea, California, USA. *Dis. Aquat. Org.* **1999**, *39*, 65–73. [CrossRef] [PubMed]
6. Shinn, A.P.; Pratoomyot, J.; Bron, J.E.; Paladini, G.; Brooker, E.E.; Brooker, A.J. Economic costs of protistan and metazoan parasites to global mariculture. *Parasitology* **2015**, *142*, 196–270. [CrossRef]
7. Balcázar, J.L.; Blas, I.D.; Ruiz-Zarzuela, I.; Cunningham, D.; Vendrell, D.; Múzquiz, J.L. The role of probiotics in aquaculture. *Vet. Microbiol.* **2006**, *114*, 173–186. [CrossRef]
8. Mladineo, I. Check list of the parasitofauna in Adriatic Sea cage-reared fish. *Acta Vet.* **2006**, *56*, 285–292. [CrossRef]
9. Lawler, A.R. Studies on *Amyloodinium ocellatum* (Dinoflagellata) in Mississippi Sound: Natural and Experimental Hosts. *Gulf Res. Rep.* **1980**, *6*, 403–413. [CrossRef]
10. Paperna, I. Chemical control of *Amyloodinium ocellatum* (Brown 1931) (Dinoflagellida) infections: In vitro tests and treatment trials with infected fishes. *Aquaculture* **1984**, *38*, 1–18. [CrossRef]
11. Fajer-Avila, E.J.; Abdo-de la Parra, I.; Aguilar-Zarate, G.; Contreras-Arce, R.; Zaldivar-Ramirez, J.; Betancourt-Lozano, M. Toxicity of formalin to bullseye puffer fish (*Sphoeroides annulatus* Jenyns, 1843) and its effectiveness to control ectoparasites. *Aquaculture* **2003**, *223*, 41–50. [CrossRef]

12. Kabata, Z. *Parasites and Diseases of Fish Cultured in the Tropics*; Taylor & Francis Ltd.: London, UK, 1985.
13. Virgula, J.C.; Cruz-Lacierda, E.R.; Estante, E.G.; Corre, V.L., Jr. Copper sulfate as treatment for the ectoparasite *Amyloodinium ocellatum* (Dinoflagellida) on *milkfish* (Chanos chanos) fry. *Aquac. Aquar. Conserv. Legis.* **2017**, *10*, 365–371.
14. Bessat, M.; Fadel, A. Amyloodiniosis in cultured *Dicentrarchus labrax*: Parasitological and molecular diagnosis, and an improved treatment protocol. *Dis. Aquat. Org.* **2018**, *129*, 41–51. [CrossRef]
15. Montgomery-Brock, D.; Sato, V.T.; Brock, J.A.; Tamaru, C.S. The application of hydrogen peroxide as a treatment for the ectoparasite *Amyloodinium ocellatum* (Brown 1931) on the Pacific threadfin *Polydactylus sexfilis*. *J. World Aquac. Soc.* **2001**, *32*, 250–254. [CrossRef]
16. Seoud, S.S.M.; Zaki, V.H.; Ahmed, G.E.; El-Khalek, N.K.A. Studies on Amyloodinium Infestation in European Seabass (*Dicentrarchus labrax*.) Fishes with Special Reference for Treatment. *Int. J. Mar. Sci.* **2017**, *7*. [CrossRef]
17. Oestmann, D.J.; Lewis, D.H.; Zettler, B.A. Communications: Clearance of *Amyloodinium ocellatum* Dinospores by *Artemia salina*. *J. Aquat. Anim. Health* **1995**, *7*, 257–261. [CrossRef]
18. Tahraoui, S.; Ennaffah, B.; Belattmania, Z.; Reani, A.; Sabour, B. First Report on the Occurrence and Dynamics of the Ectoparasitic Dinoflagellate *Amyloodinium ocellatum* in the Moroccan Atlantic Coast. *Res. J. Environ. Sci.* **2018**, *12*, 153–159.
19. Gómez, F.; Gast, R.J. Dinoflagellates Amyloodinium and Ichthyodinium (Dinophyceae), parasites of marine fishes in the South Atlantic Ocean. *Dis. Aquat. Org.* **2018**, *131*, 29–37. [CrossRef]
20. Raposo de Magalhães, C.S.F.; Cerqueira, M.A.C.; Schrama, D.; Moreira, M.J.V.; Boonanuntanasarn, S.; Rodrigues, P.M.L. A Proteomics and other Omics approach in the context of farmed fish welfare and biomarker discovery. *Rev. Aquac.* **2018**. [CrossRef]
21. Conte, F.S. Stress and the welfare of cultured fish. *Appl. Anim. Behav. Sci.* **2004**, *86*, 205–223. [CrossRef]
22. Pavlidis, M.; Angellotti, L.; Papandroulakis, N.; Divanach, P. Evaluation of transportation procedures on water quality and fry performance in red porgy (*Pagrus pagrus*) fry. *Aquaculture* **2003**, *218*, 187–202. [CrossRef]
23. Conceição, L.E.C.; Aragão, C.; Dias, J.; Costas, B.; Terova, G.; Martins, C.; Tort, L. Dietary nitrogen and fish welfare. *Fish Physiol. Biochem.* **2012**, *38*, 119–141. [CrossRef]
24. Alves, R.N.; Cordeiro, O.; Silva, T.S.; Richard, N.; de Vareilles, M.; Marino, G.; Di Marco, P.; Rodrigues, P.M.; Conceição, L.E.C. Metabolic molecular indicators of chronic stress in gilthead seabream (*Sparus aurata*) using comparative proteomics. *Aquaculture* **2010**, *299*, 57–66. [CrossRef]
25. Cordeiro, O.; Silva, T.; Alves, R.; Costas, B.; Wulff, T.; Richard, N.; de Vareilles, M.; Conceição, L.C.; Rodrigues, P. Changes in Liver Proteome Expression of Senegalese Sole (*Solea senegalensis*) in Response to Repeated Handling Stress. *Mar. Biotechnol.* **2012**, *14*, 714–729. [CrossRef]
26. Baldwin, L. The effects of stocking density on fish welfare. *Plymouth Stud. Sci.* **2011**, *4*, 372–383.
27. Ashley, P.J. Fish welfare: Current issues in aquaculture. *Appl. Anim. Behav. Sci.* **2007**, *104*, 199–235. [CrossRef]
28. Wall, T. Disease and Medicines—The Welfare Implications. In *Fish Welfare*; Blackwell Publishing Ltd.: Oxford, UK, 2008; pp. 195–201.
29. Einarsdóttir, I.E.; Nilssen, K.J.; Iversen, M. Effects of rearing stress on Atlantic salmon (*Salmo salar* L.) antibody response to a non-pathogenic antigen. *Aquac. Res.* **2000**, *31*, 923–930. [CrossRef]
30. Paperna, I. Reproduction cycle and tolerance to temperature and salinity of *Amyloodinium ocellatum* (Brown, 1931) (Dinoflagellida). *Ann. Parasitol. Hum. Comp.* **1984**, *59*, 7–30. [CrossRef]
31. Noga, E.J. *Amyloodinium ocellatum*. In *Fish Parasites: Pathobiology and Protection*; Woo, P.T.K., Buchmann, K., Eds.; CABI Publishers: Preston, UK, 2012; pp. 19–29.
32. Kuperman, B.I.; Matey, V.E.; Hurlbert, S.H. Parasites of fish from the Salton Sea, California, USA. *Hydrobiologia* **2001**, *466*, 195–208. [CrossRef]
33. Pereira, J.C.; Abrantes, I.; Martins, I.; Barata, J.; Frias, P.; Pereira, I. Ecological and morphological features of *Amyloodinium ocellatum* occurrences in cultivated gilthead seabream *Sparus aurata* L.; A case study. *Aquaculture* **2011**, *310*, 289–297. [CrossRef]
34. Noga, E.J.; Fan, Z.; Silphaduang, U. Host site of activity and cytological effects of histone-like proteins on the parasitic dinoflagellate *Amyloodinium ocellatum*. *Dis. Aquat. Organ.* **2002**, *52*, 207–215. [CrossRef]
35. Colorni, A.; Ullal, A.; Heinisch, G.; Noga, E.J. Activity of the antimicrobial polypeptide piscidin 2 against fish ectoparasites. *J. Fish Dis.* **2008**, *31*, 423–432. [CrossRef]
36. Woo, P.T. Protective immunity in fish against protozoan diseases. *Parassitologia* **2007**, *49*, 185–191. [PubMed]

37. Alvarez-Pellitero, P. Fish immunity and parasite infections: From innate immunity to immunoprophylactic prospects. *Vet. Immunol. Immunopathol.* **2008**, *126*, 171–198. [CrossRef]
38. Byadgi, O.; Beraldo, P.; Volpatti, D.; Massimo, M.; Bulfon, C.; Galeotti, M. Expression of infection-related immune response in European sea bass (*Dicentrarchus labrax*) during a natural outbreak from a unique dinoflagellate *Amyloodinium ocellatum*. *Fish Shellfish Immunol.* **2019**, *84*, 62–72. [CrossRef]
39. Nozzi, V.; Strofaldi, S.; Piquer, I.F.; Di Crescenzo, D.; Olivotto, I.; Carnevali, O. *Amyloodinum ocellatum* in *Dicentrarchus labrax*: Study of infection in salt water and freshwater aquaponics. *Fish Shellfish Immunol.* **2016**, *57*, 179–185. [CrossRef]
40. Guerra-Santos, B.; Albinati, R.C.B.; Moreira, E.L.T.; Lima, F.W.M.; de Azevedo, T.M.P.; Costa, D.S.P.; de Medeiros, S.D.C.; Lira, A.D. Parameters hematological and histopathologic alterations in cobia (*Rachycentron canadum* Linnaeus, 1766) com amyloodiniose. *Pesq. Vet. Bras.* **2012**, *32*, 1184–1190. [CrossRef]
41. Moreira, M.; Schrama, D.; Soares, F.; Wulff, T.; Pousão-Ferreira, P.; Rodrigues, P. Physiological responses of reared sea bream (*Sparus aurata* Linnaeus, 1758) to an *Amyloodinium ocellatum* outbreak. *J. Fish Dis.* **2017**, *40*, 1545–1560. [CrossRef] [PubMed]
42. Moreira, M.; Herrera, M.; Pousão-Ferreira, P.; Navas Triano, J.I.; Soares, F. Stress effects of amyloodiniosis in gilthead sea bream *Sparus aurata*. *Dis. Aquat. Org.* **2018**, *127*, 201–211. [CrossRef] [PubMed]
43. Vivanco-Aranda, M.; Del Río-Zaragoza, O.B.; Lechuga-Sandoval, C.E.; Viana, M.T.; Rombenso, A.N. Health response in yellowtail *Seriola dorsalis* exposed to an *Amyloodinium ocellatum* outbreak. *Cienc. Mar.* **2018**, *44*. [CrossRef]
44. Reyes-Becerril, M.; Ascencio-Valle, F.; Alamillo, E.; Hirono, I.; Kondo, H.; Jirapongpairoj, W.; Angulo, C. Molecular cloning and comparative responses of Toll-like receptor 22 following ligands stimulation and parasitic infection in yellowtail (*Seriola lalandi*). *Fish Shellfish Immunol.* **2015**, *46*, 323–333. [CrossRef] [PubMed]
45. Cheung, P.J.; Nigrelli, R.F.; Ruggieri, G.D. *Oodinium ocellatum* (Brown, 1931) (Dinoflagellata) in the kidney and other internal tissues of pork fish, *Anisotremus virginicus* (L.). *J. Fish Dis.* **1981**, *4*, 523–525. [CrossRef]
46. Yemmen, C.; Bahri, S. Seasonality of *Amyloodinium ocellatum* Brown 1931 (Dinophyceae) infesting the Senegalese sole *Solea senegalensis* from Bizerte lagoon, Tunisia. *J. Fish Dis.* **2017**, *40*, 853–857. [CrossRef]
47. Soares, F.; Quental-Ferreira, H.; Moreira, M.; Cunha, E.; Ribeiro, L.; Pousao-Ferreira, P. First report of *Amyloodinium ocellatum* in farmed meagre (*Argyrosomus regius*). *Bull. Eur. Assoc. Fish Pathol.* **2012**, *32*, 30–33.
48. Kumar, P.R.; Nazar, A.K.A.; Jayakumar, R.; Tamilmani, G.; Sakthivel, M.; Kalidas, C.; Balamurugan, V.; Sirajudeen, S.; Thiagu, R.; Gopakumar, G. *Amyloodinium ocellatum* infestation in the broodstock of silver pompano *Trachinotus blochii* (Lacepede, 1801) and its therapeutic control. *Indian J. Fish.* **2015**, *62*, 131–134.
49. Ángeles Esteban, M. An Overview of the Immunological Defenses in Fish Skin. *ISRN Immunol.* **2012**, *2012*, 853470. [CrossRef]
50. Guardiola, F.A.; Cuesta, A.; Abellán, E.; Meseguer, J.; Esteban, M.A. Comparative analysis of the humoral immunity of skin mucus from several marine teleost fish. *Fish Shellfish Immunol.* **2014**, *40*, 24–31. [CrossRef] [PubMed]
51. Paperna, I. Amyloodinium-Ocellatum (Brown, 1931) (Dinoflagellida) Infestations in Cultured Marine Fish at Eilat, Red-Sea—Epizootiology and Pathology. *J. Fish Dis.* **1980**, *3*, 363–372. [CrossRef]
52. Triki, Z.; Grutter, A.S.; Bshary, R.; Ros, A.F.H. Effects of short-term exposure to ectoparasites on fish cortisol and hematocrit levels. *Mar. Biol.* **2016**, *163*, 187. [CrossRef]
53. Olivotto, I.; Mosconi, G.; Maradonna, F.; Cardinali, M.; Carnevali, O. *Diplodus sargus* interrenal–pituitary response: Chemical communication in stressed fish. *Gen. Comp. Endocrinol.* **2002**, *127*, 66–70. [CrossRef]
54. Olsen, R.E.; Sundell, K.; Ringø, E.; Myklebust, R.; Hemre, G.-I.; Hansen, T.; Karlsen, Ø. The acute stress response in fed and food deprived Atlantic cod, *Gadus morhua* L. *Aquaculture* **2008**, *280*, 232–241. [CrossRef]
55. Magalhães, R.; Martins, N.; Martins, S.; Lopes, T.; Díaz-Rosales, P.; Pousão-Ferreira, P.; Oliva-Teles, A.; Peres, H. Is dietary taurine required for white seabream (*Diplodus sargus*) juveniles? *Aquaculture* **2019**, *502*, 296–302. [CrossRef]
56. Enes, P.; Peres, H.; Pousão-Ferreira, P.; Sanchez-Gurmaches, J.; Navarro, I.; Gutiérrez, J.; Oliva-Teles, A. Glycemic and insulin responses in white sea bream *Diplodus sargus*, after intraperitoneal administration of glucose. *Fish Physiol. Biochem.* **2012**, *38*, 645–652. [CrossRef]
57. González Gómez, M.P.; Marín Arribas, S.L.; Vargas-Chacoff, L. Stress response of *Salmo salar* (Linnaeus 1758) facing low abundance infestation of *Caligus rogercresseyi* (Boxshall & Bravo 2000), an object in the tank, and handling. *J. Fish Dis.* **2016**, *39*, 853–865. [CrossRef]

58. Bayne, C.J.; Gerwick, L. The acute phase response and innate immunity of fish. *Dev. Comp. Immunol.* **2001**, *25*, 725–743. [CrossRef]
59. Alvarez-Pellitero, P.; Palenzuela, O.; Sitjà-Bobadilla, A. Histopathology and cellular response in *Enteromyxum leei* (Myxozoa) infections of *Diplodus puntazzo* (Teleostei). *Parasitol. Int.* **2008**, *57*, 110–120. [CrossRef]
60. Horton, T.; Okamura, B. Post-haemorrhagic anaemia in sea bass, *Dicentrarchus labrax* (L.), caused by blood feeding of *Ceratothoa oestroides* (Isopoda: Cymothoidae). *J. Fish Dis.* **2003**, *26*, 401–406. [CrossRef]
61. Portz, D.E.; Woodley, C.M.; Cech, J.J. Stress-associated impacts of short-term holding on fishes. *Rev. Fish Biol. Fish.* **2006**, *16*, 125–170. [CrossRef]
62. Fallah, F.J.; Khara, H.; Rohi, J.D.; Sayadborani, M. Hematological parameters associated with parasitism in pike, *Esox lucius* caught from Anzali wetland. *J. Parasit. Dis. Off. Organ Indian Soc. Parasitol.* **2015**, *39*, 245–248. [CrossRef] [PubMed]
63. Pousão-Ferreira, P.; Gonçalves, C.; Dores, E. *Larval Rearing of Four Sparidae Species*; Special Publication No. 36; European Aquaculture Society: Oostende, Belgium, 2005.
64. Barata, M.; Soares, F.; Aragão, C.; Almeida, A.C.; Pousão-Ferreira, P.; Ribeiro, L. Efficiency of 2-phenoxyethanol and Clove Oil for Reducing Handling Stress in Reared Meagre, *Argyrosomus regius* (Pisces: Sciaenidae). *J. World Aquac. Soc.* **2016**, *47*, 82–92. [CrossRef]
65. Martoja, R.; Martoja-Pierson, M. *Initiation aux techniques de l'histologie animale*; Masson: Paris, France, 1967.
66. Herrera, M.; Ruiz-Jarabo, I.; Vargas-Chacoff, L.; de la Roca, E.; Mancera, J.M. Metabolic enzyme activities in relation to crowding stress in the wedge sole (*Dicologoglossa cuneata*). *Aquac. Res.* **2015**, *46*, 2808–2818. [CrossRef]
67. Soares, F.; Leitão, A.; Moreira, M.; de Sousa, J.T.; Almeida, A.C.; Barata, M.; Feist, S.W.; Pousão-Ferreira, P.; Ribeiro, L. Sarcoma in the thymus of juvenile meagre *Argyrosomus regius* reared in an intensive system. *Dis. Aquat. Organ.* **2012**, *102*, 119–127. [CrossRef] [PubMed]
68. Matias, A.C.; Ribeiro, L.; Araujo, R.L.; Pousão-Ferreira, P. Preliminary studies on haematological and plasmatic parameters in gilthead sea bream (*Sparus aurata*) held under day/night temperature variations. *Fish Physiol. Biochem.* **2017**, *44*, 273–282. [CrossRef] [PubMed]
69. Radu, D.; Oprea, L.; Bucur, C.; Costache, M.; Oprea, D. Characteristics of haematological parameters for carp culture and Koi (*Cyprinus carpio* Linneaus, 1758) reared in an intensive system. *Bull. Uasvm Anim. Sci. Biotechnol.* **2009**, *66*, 336–342.

*fishes*

MDPI

*Article*

# Enrichment Increases Aggression in Zebrafish

**Melanie A. Woodward, Lucy A. Winder and Penelope J. Watt \***

Department of Animal and Plant Sciences, University of Sheffield, Sheffield S10 2TN, UK;
mel_woodward1203@hotmail.com (M.A.W.); lwinder1@sheffield.ac.uk (L.A.W.)
\* Correspondence: p.j.watt@sheffield.ac.uk; Tel.: +44-114-2220076

Received: 20 December 2018; Accepted: 12 March 2019; Published: 19 March 2019

check for
updates

**Abstract:** Environmental enrichment, or the enhancement of an animal's surroundings when in captivity to maximise its wellbeing, has been increasingly applied to fish species, particularly those used commercially. Laboratory species could also benefit from enrichment, but it is not always clear what constitutes an enriched environment. The zebrafish, *Danio rerio*, is used widely in research and is one of the most commonly-used laboratory animals. We investigated whether changing the structural complexity of housing tanks altered the behaviour of one strain of zebrafish. Fish were kept in three treatments: (1) very enhanced (VE); (2) mildly enhanced (ME); and (3) control (CT). Level of aggression, fertilisation success, and growth were measured at regular intervals in a subset of fish in each treatment group. The VE fish were more aggressive over time than either ME or CT fish, both in the number of attacks they made against a mirror image and in their tendency to stay close to their reflection rather than avoid it. Furthermore, VE fish were shorter than CT fish by the end of the experiment, though mass was not significantly affected. There was no significant effect of treatment on fertilisation success. These findings suggest that the way in which fish are housed in the laboratory can significantly affect their behaviour, and potentially, their growth. The zebrafish is a shoaling species with a dominance hierarchy, and so may become territorial over objects placed in the tank. The enrichment of laboratory tanks should consider aspects of the species' behaviour.

**Keywords:** *Danio rerio*; structural complexity; aggression; territorial; boldness; fertilisation success

## 1. Introduction

Laboratory animal holding facilities conventionally have been designed on the basis of human requirements, both economic and ergonomic. Therefore, the captive conditions in which animals are kept are vastly different from their natural environment, and so the behaviour of such animals is likely to be unnatural. Indeed, captivity could adversely affect the behaviour and physiology of animals, for example, if their blood chemistry and metabolism are altered due to unnatural stress levels [1]. In an attempt to improve scientific validity, laboratories are beginning to alter housing facilities in order to simulate more natural environments, a concept known as environmental enrichment [2].

Environmental enrichment has traditionally focussed on mammalian species, specifically domestic and farm animals [1]. Although a variety of definitions exist, the emphasis is generally on increasing the amount of stimulation the captive environment provides for the animals [1]. Smith and Taylor [2] state that environmental enrichment should allow an animal to express a more natural behavioural repertoire. However, this highlights a significant problem: how to define the natural or normal behaviour of an animal [3]?

Despite fish being commonly used in laboratories [1], information about how to make them behave naturally or live stress-free lives could be improved. Environmental enrichment has been widely studied in commercially reared species of fish in an attempt to boost populations when they are released into the wild (e.g., [4–6]). However, less is known about the environmental and biological requirements of cultured species [7,8], and so improving the current knowledge of the impact of environmental enrichment on different fish species is important (see Reference [9] for a review of this subject).

Environmental enrichment within both commercial and laboratory aquaria often involves increasing the structural complexity of the tank by adding various objects to simulate natural habitats, including gravel, stones, real and artificial plants, plastic tubing, driftwood and upturned plant pots (e.g., [10–12]), and so reducing abnormal traits [9]. Another common technique used in fish studies involves altering tank lighting or colour (e.g., [13–19]) in an attempt to simulate a more natural environment.

Previous studies that compared fish reared in bare aquaria to those in enriched tanks found significant behavioural differences between the two. For example, fish in structured environments have been found to display reduced aggression levels (zebrafish, *Danio rerio* [20,21], salmonids [22]), possibly due to the structures reducing the visibility between individuals and so reducing aggressive interactions. Several studies have found that fish kept in bare tanks are more exploratory than those in tanks with structures added, which are more cautious in a novel environment (zebrafish [12], Atlantic salmon, *Salmo salar* [23]). However, Roberts et al. [23] found that when conspecific alarm cues were released into the water, Atlantic salmon housed in structurally complex tanks took significantly less time to leave a shelter.

There are mixed findings regarding the growth of fish in enriched and bare tanks, as some studies have found that fish kept in bare tanks are larger (Chinook salmon, *Oncorhynchus tshawytscha* [11], zebrafish [12]), whereas others show enriched fish to be larger (common carp, *Cyprinus carpio* [14], thinlip mullet, *Liza ramada* [19], Atlantic salmon [24]).

The zebrafish is an important vertebrate model organism in a variety of biological disciplines and is a commonly-used laboratory animal. It has been suggested that enrichment could improve the quality of care for zebrafish [25] and they seem to prefer structure in a tank [26], but there is a lack of specific guidelines regarding enrichment for this species. Indeed, plants, which are commonly used to enrich tanks, can have both positive and negative effects on fish (reduced forebrain, Chinook salmon [11], slower rate of learning, zebrafish [12], reduced aggression, zebrafish [20,21], reduced risk taking, Atlantic salmon [23]).

In this study, we investigated the effect of various types of tank enhancement, feasible in a laboratory setting, on the behaviour (aggression, boldness, activity), reproduction and growth of one strain of zebrafish by manipulating the tank environment. The results showed that aggression can increase in fish from tanks that have a higher structural diversity, and that growth may be affected.

## 2. Results

### 2.1. Aggression

The type of tank enhancement that fish were kept in significantly affected the mean number of aggressive attacks fish made towards their image in the mirror test over the course of the experiment ($R^2 = 0.40$, $p = 0.034$; Figure 1), with fish from very enhanced (VE) tanks making significantly more attacks than fish from both control (CT) and mildly enhanced (ME) tanks, given the week (CT: $p = 0.017$, ME: $p = 0.036$). Week affected the number of attacks differently over time for each enhancement treatment. VE tanks promoted more attacks over time, whereas CT and ME tanks had slightly fewer attacks over time (model estimates in comparison to VE; CT: −0.05, ME: −0.04).

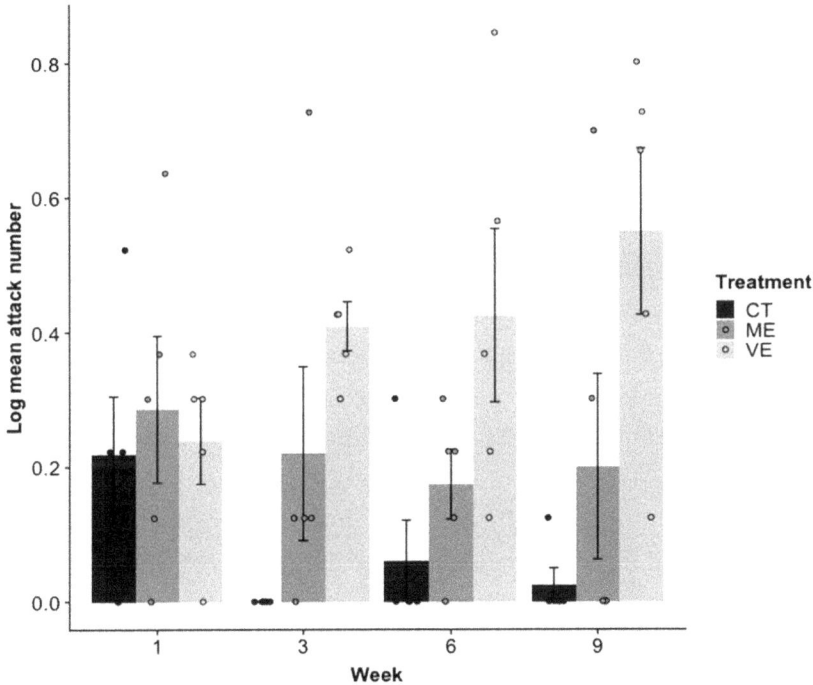

**Figure 1.** Mean ± SE number of aggressive attacks made towards a mirror image in the mirror test in control (CT), mildly enhanced (ME), and very enhanced (VE) treatments in weeks 1, 3, 6, and 9. There were five replicates of each treatment in each week (each replicate is shown as a point).

There was a significant interaction between section, treatment, and week for the mean number of times that fish entered a section ($R^2$ = 0.55, $p$ = 0.002; Figure 2). Fish preferred sections B and C throughout the experiment for CT and ME tanks, which suggested they were neither avoiding nor interacting with the mirror image. This was also true of fish in the VE tanks for weeks 1 and 3. However, in week 6, fish spent more time in sections A and B than C and D. This suggests that fish in the VE tank began to interact with the mirror. In week 9, fish in the VE tank entered each section approximately the same number of times.

The number of aggressive attacks conducted by fish was significantly correlated with the time they spent closest to their mirror image (section A) throughout the experiment (Pearson correlation, $r$ = 0.27, df = 58, $p$ < 0.05), suggesting that section preference was an indication of aggressiveness.

*2.2. Boldness*

Boldness, the propensity of individuals to explore a novel environment, was not significantly affected by type of tank enhancement ($p$ = 0.85) or week ($p$ = 0.43). The mean freezing time, when fish were completely still and often associated with anxiety, did not differ significantly among enhancement treatments ($p$ = 0.58) or between weeks ($p$ = 0.45).

*2.3. Activity Level*

Activity level, the degree to which fish actively moved in their home tank, was not significantly affected by type of tank enhancement ($p$ = 0.98) or week ($p$ = 0.75). Overall, activity level (Mean ± SE: 88.3 ± 2.1) was significantly lower than boldness (246.6 ± 7.3) (Two-sample *t*-test: $t$ = 20.75, df = 118, $p$ < 0.001).

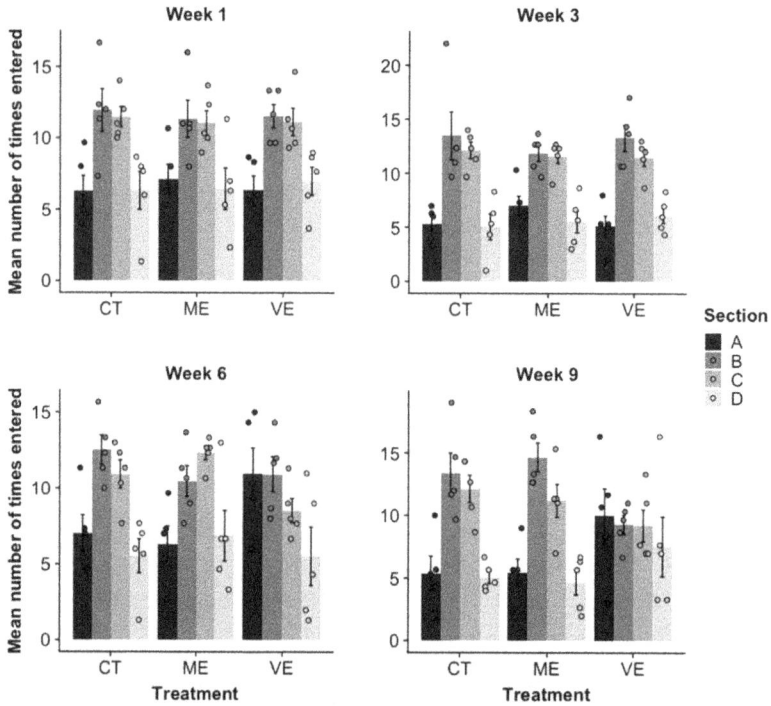

**Figure 2.** Mean ± SE number of times fish entered sections A, B, C, and D in the mirror test in control (CT), mildly enhanced (ME), and very enhanced (VE) tank treatments in weeks 1, 3, 6, and 9. Section A was closest to the mirror, section D was furthest from the mirror, and sections B and C were in the middle of the tank. There were five replicates per treatment.

*2.4. Reproduction*

The mean number of eggs produced by females was significantly affected by the type of tank enhancement ($R^2$ = 0.32, $p$ = 0.004; Figure 3). Mild enhancement (ME) saw females producing 54% more eggs than control tanks (CT) and 81% more than very enhanced (VE) (CT: $p$ = 0.05; VE: $p$ = 0.001). The mean number of eggs produced in CT and VE tanks was not significantly different from each other ($p$ = 0.32). Week also significantly affected the mean number of eggs produced ($p$ = 0.003; Figure 3). Fewer eggs were produced in Week 1 than in Weeks 3 and 6 (Week 3: estimate = 0.83, $p$ = 0.01; Week 6: estimate = 1.11, $p$ < 0.001). However, the mean number of eggs was not significantly different between Weeks 1 and 9 ($p$ = 0.24).

The proportion of fertilised eggs differed significantly between weeks ($R^2$ = 0.22, $p$ = 0.01; Figure 4). Fertilisation was significantly lower in Week 1 than Week 3 ($p$ = 0.001), but was not significantly different from Week 6 ($p$ = 0.09) or Week 9 ($p$ = 0.32). Week 3 had a higher proportion of eggs fertilised than all other weeks ($p$ < 0.001). There was no significant difference in the proportion of eggs fertilised between treatments ($p$ = 0.57).

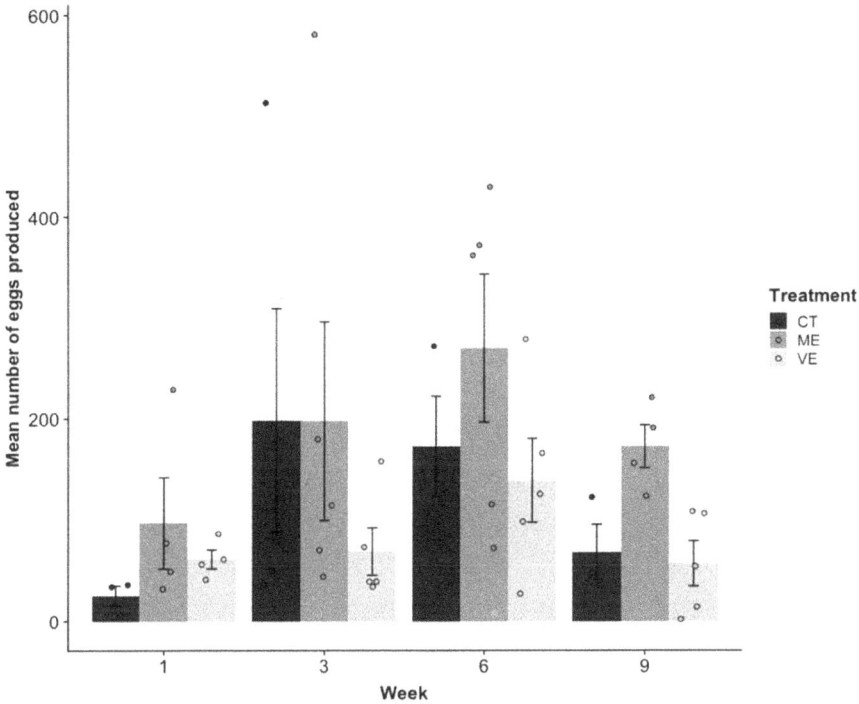

**Figure 3.** Mean ± SE number of eggs produced from the control (CT), mildly enhanced (ME), and very enhanced (VE) tanks over weeks 1, 3, 6, and 9. Sample sizes varied between 3 and 5 tanks, depending on the number of tanks that produced eggs. There was a significant difference in mean number of eggs between treatments in week 9 only, where more eggs were produced in ME tanks.

## 2.5. Growth

At the start of the experiment, the mean mass of fish differed significantly between the different enhancement tanks (Mean ± SE: CT 0.487 ± 0.008; ME 0.514 ± 0.010; VE 0.462 ± 0.008; ANOVA: $F_{2,12} = 8.92$, $p < 0.01$) because the fish randomly placed into the ME tanks were significantly heavier than those placed into the VE tanks (Tukey's test: $p < 0.01$). This pattern remained consistent at the end of the experiment in Week 9. Again, enhancement treatment had a significant effect on mean mass (Mean ± SE: CT 0.484 ± 0.017; ME 0.505 ± 0.011; VE 0.445 ± 0.009; ANOVA: $F_{2,12} = 5.90$, $p < 0.05$), with ME fish being significantly heavier than VE fish (Tukey's test: $p < 0.05$), and there was no significant difference between fish in CT tanks and either ME tanks (Tukey's test: $p = 0.49$) or VE tanks (Tukey's test: $p = 0.11$). There was no significant change in mass between the start and the end for fish in any treatment (CT: Paired *t*-test: $t = 0.21$, $df = 4$, $p = 0.84$; ME: Paired *t*-test: $t = 1.25$, $df = 4$, $p = 0.28$; VE: Paired *t*-test: $t = 1.35$, $df = 4$, $p = 0.25$).

At the start of the experiment, there was no significant difference in the mean length of the fish allocated to each enhancement type (Mean ± SE: CT 30.459 ± 0.160; ME 30.480 ± 0.178; VE 30.030 ± 0.233; ANOVA: F = 1.74, $df = 2$, $p = 0.22$). However, at the end of the experiment, in Week 9, there was a significant effect of enhancement treatment on mean length (Mean ± SE: CT 30.526 ± 0.341; ME 30.138 ± 0.255; VE 29.436 ± 0.105; ANOVA: F = 4.76, $df = 2$, $p < 0.05$) and fish housed in VE tanks were significantly shorter than fish in CT tanks (Tukey's test: $p < 0.05$).

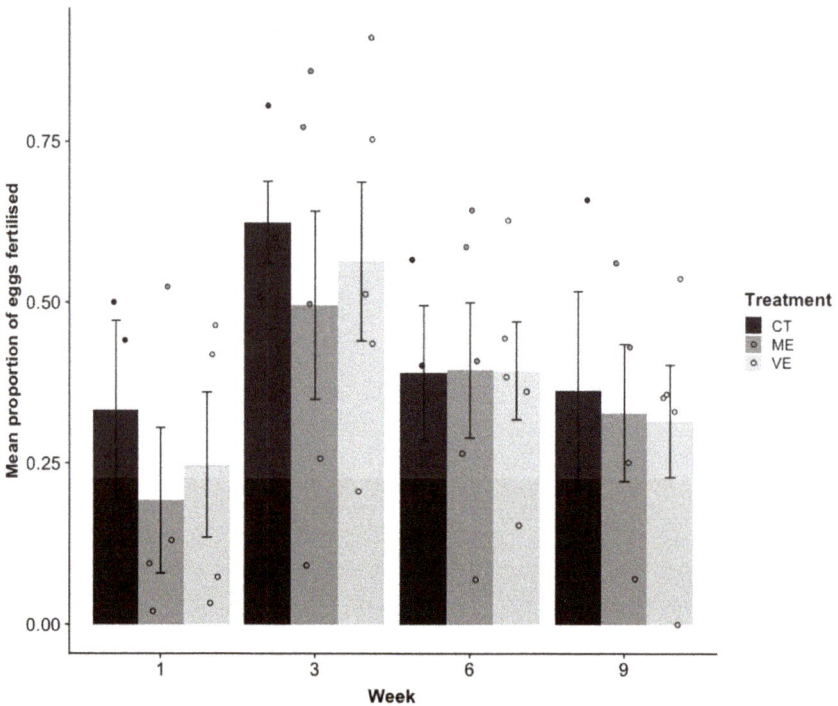

**Figure 4.** Mean ± SE proportion of eggs fertilised from the control (CT), mildly enhanced (ME), and very enhanced (VE) tanks in weeks 1, 3, 6, and 9. Sample sizes vary between 3 and 5 tanks depending on the number of tanks that produced eggs.

## 3. Discussion

Overall, our results show that the tank environment can significantly affect the behaviour, and potentially, the growth of zebrafish, at least in the strain studied. Aggression levels, both in terms of the number of aggressive attacks made and the time spent close to their mirror image, were significantly increased in fish from VE tanks compared to fish from the less (ME) or non-enhanced (CT) treatments. This increase in aggression towards their mirror image was seen from Week 3 of the experiment onwards, and time spent close to the mirror was found to correlate with the number of aggressive attacks throughout the experiment. We did not measure aggression in the tanks themselves, so it is possible that fish behaved differently when tested in front of a mirror in isolation.

In contrast, previous studies have found a decrease in aggression with structural complexity both in a laboratory setting (crayfish, *Orconectes propinquus* [27], zebrafish [20,21]) and in semi-natural environments (white-spotted char, *Salvelinus leucomaenis* [22]). Both Basquill and Grant [20] and Carfagnini et al. [21] found that zebrafish aggression rate was significantly reduced when measured in structurally complex versus bare environments, though the densities that Basquill and Grant [20] used were very low, so the fish may have been visually obscured by structures in the tank. Furthermore, both Carfagnini et al. [21] and Basquill and Grant [20] used juvenile, rather than adult zebrafish, which could suggest that life stage is important. Hamilton and Dill [28] found no difference in the aggression levels of adult zebrafish in a vegetated and a non-vegetated habitat. Our experiments ran for nine weeks and it was only after three weeks that aggression was seen to increase, suggesting that there were long-term effects on aggression when structure was present. Furthermore, as mentioned already, we measured aggression under controlled conditions, whereas in these other studies it was measured in the experimental tanks.

A possible explanation for the increased aggression in the VE fish in our study is that the fish become territorial over the objects within the tank and so their aggression level increased when measured with the mirror test. The zebrafish is a shoaling species, known to develop dominance hierarchies [29–31]. Previous studies have shown that zebrafish are highly territorial over spawning sites [32] and will defend food sources [28,33]. The structure within the VE tanks could have represented potential spawning sites, though there was no difference in the number of eggs collected in this treatment, which might be expected if fish were spawning in alternative sites in some of the treatments. Alternatively, VE fish may have become territorial over the upturned plant pots and artificial plants because food often became trapped under or around them, and so the fish could have been more aggressive when there was more structure because they were defending this resource. McCarthy et al. [34] found that in rainbow trout (*Oncorhynchus mykiss*), territories were very easy to defend when the density was low, but almost impossible to defend at high densities, and so aggressiveness over potential territories could vary with density, as found in the damselfish, *Stegastes partitus* [35,36], with aggression highest at intermediate densities like those used here. In our study, a random sample of fish was chosen from each tank for the mirror test and an equal sex ratio of fish was maintained in each tank, making it highly likely that the increased aggression observed was in both sexes, unless one sex was easier to catch than the other and so was always tested, but we have no evidence for this.

Increased plasma cortisol levels, known to be associated with stress in fish (e.g., salmonids [37]), have been demonstrated in fish displaying aggressive interactions during the development of territories and dominance hierarchies (cichlid, *Haplochromis burtoni* [38], rainbow trout [39]). Wilkes et al. [40] found no difference in cortisol levels or aggression between juvenile zebrafish kept in structured tanks and those in bare tanks. Pairs of zebrafish were found to show increased cortisol levels initially when housed with vegetation, but this declined with time, so by day 10 fish had lower levels than those kept individually or in pairs without vegetation [41]. Cortisol was not measured in our experiments, but the increased aggression within VE tanks is likely to have a negative impact on fish physiology and welfare, in which case enhancement of housing as implemented here may be detrimental, rather than beneficial, as usually assumed [20–22]. However, some degree of aggression may not be completely negative, according to the concept of allostasis [42]. The effect of enrichment on aggression may be age- and species-specific, and dependent on the territoriality of the species in question.

Fish housed in the VE tanks were significantly shorter than those in the CT tanks by the end of the experiment, and a significant change in length of these fish was detected, though no comparable change in mass was detected. This suggests there may have been an effect of treatment on growth rate, but the results are not conclusive. Previous studies have found that structure in a tank can have a positive effect on growth. For example, Finstad et al. [24] found that in Atlantic salmon, mass loss (common over winter [43]) was lowest when structural complexity was highest, possibly because there was less need for anti-predator behaviour and less competition for shelters. However, consistent with our study, others have found that fish in bare tanks have higher growth rates than those with structure (Chinook salmon [11], zebrafish [12]), though some have found no effect of structure on growth (zebrafish [44]), but an effect on morphology (three-spine stickleback, *Gasterosteus aculeatus* [45]). Kihslinger et al. [11] found that after one year of rearing, hatchery Chinook salmon were significantly heavier than their wild counterparts and similar results have been seen in the laboratory. Spence et al. [12] found that zebrafish reared in structurally complex environments were significantly shorter than those from bare tanks. A suggested explanation for reduced growth of fish subjected to structural complexity is that the objects within the tank trap food, which is consequently less accessible to the fish, as was observed in our study. Alternatively, an increase in aggression and a decrease in growth could be due to a trade-off in energy balance.

We did not find an increase in fertilisation success with increasing enhancement. While ME tanks produced more eggs than either of the other two housing treatments, the proportion of fertilised eggs did not differ. Carfagnini et al. [21] housed zebrafish in structurally complex and bare tanks and found that fecundity was not affected by housing complexity. However, they did find that when levels of aggression were highest, fecundity decreased [21], possibly due to elevated metabolic costs associated with high aggression levels.

We found no significant difference in the boldness of fish from the different treatments, a result in common with other studies on guppies (*Poecilia reticulata*) and three-spine sticklebacks [46,47]. There has been considerable debate about the measurement of boldness and whether it could simply be a reflection of the animal's general activity level [48] but this has been found not to be the case [49]. In this study, fish were found to cross significantly more lines in the open-field test than they did in their home tanks but there was no difference between treatments. In contrast, von Krogh et al. [44] found that zebrafish from bare tanks had higher locomotor activity than those from structurally complex tanks.

The current study found no significant effect of tank background (ME tanks) on the behaviour of zebrafish, and so it appears that structural complexity within the tank and not tank background alone affected the fish. Previous work has found that fish housed in tanks with light-coloured backgrounds are more aggressive (Atlantic charr, *Salvelinus alpines* [50]), more exploratory (zebrafish [51]), have higher growth rates (common carp [14], thinlip mullet [19]) and have a higher oxygen consumption (scaled carp, *Cyprinus carpio* [13]). However, no significant effect of tank background was found in pot-bellied seahorses (*Hippocampus abdominalis*) [15] and Atlantic cod, *Gadus morhua* [17], suggesting that the effect of tank colour or background may be species-specific. It could also be dependent on the variability of the light conditions in which species live naturally, as the more variable they are, the less likely it is that differences in tank colour will have an effect.

In conclusion, this study has found that the enhancement of fish tanks, by adding structure, can significantly affect the aggressive behaviour of adult zebrafish, at least in this strain and when tested in isolation, and, potentially, their growth. Laboratories need to consider species-specific behaviour before enrichment is implemented.

## 4. Methods

### 4.1. Fish, Treatments and Growth

All fish used in the experiments were adult Nacre zebrafish. Stock fish were housed in 10-litre tanks (30 × 15 × 24 cm) in a recirculating system. Tanks were kept at 26 ± 1 °C under a 12:12 h light:dark photoperiod with 40 min dusk to dawn simulations. Fish were fed twice a day with ground-up flake food and day-old brine shrimp (*Artemia salina*).

Fish were housed in tanks of one of three treatment groups: (1) very enhanced (VE), (2) mildly enhanced (ME) or (3) control (CT) (Figure 5). Very enhanced tanks contained one artificial plant and one upturned plant pot (7 cm diameter). Additionally, plastic boards were backed with an aquarium backing paper (a blue seascape design) and placed against the rear walls of these tanks. Mildly enhanced tanks contained only the seascape boards against the rear walls, and CT tanks were left completely bare. These definitions were used purely to describe the experimental treatments and were not used as value judgements. Ten size-matched fish (five males, five females) were randomly placed in each treatment tank and there were five replicate tanks of each treatment group.

Fish were given a nine-day acclimatisation period, after which behavioural measurements were taken. The behavioural tests were done on a subset of three randomly-chosen fish per tank, and fecundity and fertilisation success were measured per tank. These measurements were taken at regular intervals throughout the course of the experiment at weeks 1, 3, 6, and 9. Standard length (tip of snout to caudal peduncle) and mass of each fish was measured at the start and end of the experiment in weeks one and nine. Previous work has shown that boldness and aggression are not related to size in either sex in this strain of zebrafish [49]. Lighting during behavioural trials was provided by two 18 W daylight fluorescent tubes placed approximately 40 cm above the test tank. Test tanks were filled with water from the recirculating system kept at the same condition as water within the housing tank system. Fish were tested before feeding in order to standardise this.

(a)

(b)

(c)

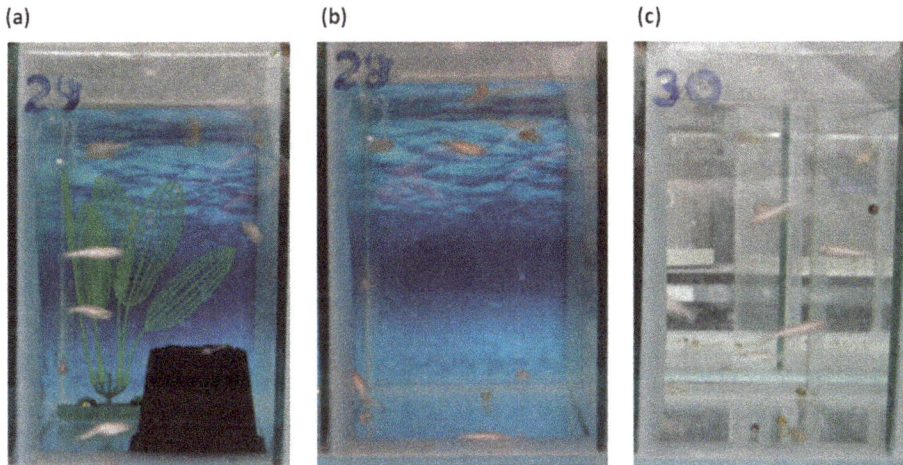

**Figure 5.** The three housing treatments: (**a**) very enhanced (VE), (**b**) mildly enhanced (ME), and (**c**) control (CT).

### 4.2. Behavioural Tests

#### 4.2.1. Aggression

Aggression was tested using the mirror test [52,53]. A test tank (28 × 17.5 × 17 cm) was filled with 4.8 L of water and a mirror (45 × 38 cm) was placed at the back of the tank at an angle of 22.5° (Figure 6). A single fish was added to the tank and allowed to acclimatise for 60 s with the mirror covered. The cover was then removed and the number of aggressive attacks that a fish performed towards its mirror image was recorded over a 180 s period. The area of the test tank in which a fish spent most of its time was also quantified immediately after the aggressive attacks were recorded. The front of the test tank was marked into four equal vertical sections (Figure 6), and the section in which the fish was located was recorded every 5 s for 180 s. A fish that spent much of its time in section A could be regarded as aggressive because it was closer to its image, whereas a fish in section D may be avoiding aggression.

#### 4.2.2. Boldness

To test for boldness, the open-field test [54–56] was used. A single fish was added to the test tank (48 × 23 × 26.5 cm), which had a 4 × 6 grid on the bottom and was filled with 3.25 L of water; the fish was left to acclimatise for 60 s. Immediately after this period, the number of lines the fish crossed, taken as a measure of boldness [49], and the total freezing time (time spent completely still) within a 180 s period, were recorded. The sides of the test tank were covered with opaque paper to minimise disturbance to the fish caused by the close proximity of the observer.

#### 4.2.3. Activity Level

Activity level was measured in home tanks by placing a 2 × 5 cm grid onto the front of the tank and recording the movements of a single fish over a period of 180 s. The number of lines the fish crossed in this period was counted. Three randomly chosen fish per tank were tested.

#### 4.2.4. Reproduction

A spawning dish (12.5 cm in diameter) containing two layers of glass marbles was placed into each tank at 16:00 on the first day of each measurement period. When fish spawned over the marbles, the eggs dropped into the spaces, stopping fish from eating them and aiding in their collection [57,58].

Moreover, zebrafish are likely to choose to spawn in these dishes as they have been shown to prefer to spawn in a substratum that provides protection to the eggs [59]. At 10:00 the following day, all dishes were taken out of the tanks, marbles were carefully removed, and all eggs within each dish were counted and transferred to a Petri dish containing aquarium water. Egg production can be variable in zebrafish [32], and so the first introduction of a spawning dish did not always lead to egg production. Therefore, if the fish in a tank did not produce any eggs, a spawning dish was placed in the tank at 16:00 every day for one week until eggs were produced. After the eggs had been in the Petri dishes for 24 h, the number that were fertilised was counted. Fertilised eggs remain transparent after this time period, but dead and unfertilised eggs become white and opaque [60,61].

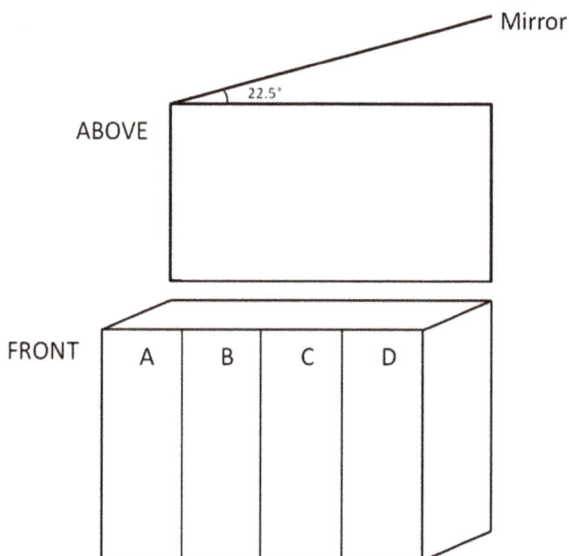

**Figure 6.** Schematic of the front and top view of the test tank used for the mirror test.

*4.3. Data Analysis*

All data analyses were conducted using the R statistical package, version 2.14.1 [62]. For all datasets, means were calculated for each tank and used in the analysis. All behavioural data were analysed with linear mixed-effects models using the nlme package (model outputs shown in Supplementary Material, Table S1). The mean attack number was log-transformed before being modelled with enhancement treatment and week as interaction terms; tank was included as a random effect in the model. The mean number of entries into a section was modelled with section, enhancement treatment, and week as a three-way interaction, with tank as a random effect. For both the mean number of attacks and number of entries, model fit was tested using the Akaike information criterion (AIC), and in both cases, no terms were dropped for either model.

Boldness was measured in two separate models: first, using the number of lines crossed within the tank as the response variable, and secondly with freeze time as the response variable. Fish activity was also modelled with the number of lines crossed in the tank as a response variable. In all the boldness and activity models, treatment and week were fixed effects, with an interaction between the two proving to be non-significant, and therefore dropped from all models, as determined by the model fit through AIC.

Reproduction was first examined with the mean number of eggs as the response variable using a negative binomial model in the "ggfortify" package, to account for over-dispersion in the data. Enhancement treatment and week were used as fixed effects and no significant interaction was found between the two. Secondly, the mean proportion of eggs fertilised was investigated using generalized linear models in the "stats" package, version 3.5.2, after being log-transformed to give normally distributed data. An interaction between enhancement treatment and week proved non-significant in explaining variation through AIC assessment of single-term deletions of fixed effects, and was therefore dropped from the model. Treatment also proved non-significant as a fixed effect and was also removed from the final model. Outputs for the reproduction models are shown in Supplementary Materials, Table S2. Paired *t*-tests were used to compare the before and after weight and length data.

This work had appropriate ethical approval from the University of Sheffield.

**Supplementary Materials:** The following are available online at http://www.mdpi.com/2410-3888/4/1/22/s1, Table S1: Outputs from behavioural generalised linear mixed-effect models of 1) aggression, 2) boldness, and 3) activity, Table S2: Model outputs of 1) egg number (mean), and 2) proportion of eggs fertilised.

**Author Contributions:** P.J.W. and M.A.W. conceived of and designed the experiments, M.A.W. conducted the work, L.A.W. and M.A.W. analysed the data, M.A.W. and P.J.W. wrote the paper.

**Funding:** This research received no external funding.

**Acknowledgments:** We thank Tolulope Ariyomo and Terry Burke for advice.

**Conflicts of Interest:** The authors declare no conflict of interest.

## References

1. Williams, T.D.; Readman, G.D.; Owen, S.F. Key issues concerning environmental enrichment for laboratory-held fish species. *Lab. Anim.* **2009**, *43*, 107–120. [CrossRef] [PubMed]
2. Smith, C.P.; Taylor, V.P. Environmental enrichment resources for laboratory animals: 1965–1995: Birds, cats, dogs, farm animals, ferrets, rabbits and rodents. *AWIC Resour. Ser.* **1995**, *2*, 145–212.
3. Newberry, R. Environmental enrichment: Increasing the biological relevance of captive environments. *Appl. Anim. Behav. Sci.* **1995**, *44*, 229–243. [CrossRef]
4. Braithwaite, V.A.; Salvanes, A.G.V. Environmental variability in the early rearing environment generates behaviourally flexible cod: Implications for rehabilitating wild populations. *Proc. R. Soc. Lond. B* **2005**, *272*, 1107–1113. [CrossRef] [PubMed]
5. Johnsson, J.I.; Brockmark, S.; Näslund, J. Environmental effects on behavioural development consequences for fitness of captive-reared fishes in the wild. *J. Fish Biol.* **2014**, *85*, 1946–1971. [CrossRef] [PubMed]
6. Rosengren, M.; Kvingedal, E.; Näslund, J.; Johnsson, J.I.; Sundell, K. Born to be wild: Effects of rearing density and environmental enrichment on stress, welfare, and smolt migration in hatchery-reared Atlantic salmon. *Can. J. Fish. Aquat. Sci.* **2017**, *74*, 396–405. [CrossRef]
7. Ellis, T.; North, B.; Scott, A.P.; Bromage, N.R.; Porter, M.; Gadd, D. The relationships between stocking density and welfare in farmed rainbow trout. *J. Fish Biol.* **2002**, *61*, 493–531. [CrossRef]
8. Ashley, P.J. Fish welfare: Current issues in aquaculture. *Appl. Anim. Behav. Sci.* **2007**, *104*, 199–235. [CrossRef]
9. Näslund, J.; Johnsson, J.I. Environmental enrichment for fish in captive environments: Effects of physical structures and substrates. *Fish Fish.* **2016**, *17*, 1–30. [CrossRef]
10. Kemp, P.S.; Armstrong, J.D.; Gilvear, D.J. Behavioural responses of juvenile Atlantic salmon (*Salmo salar*) to presence of boulders. *River Res. Appl.* **2005**, *21*, 1053–1060. [CrossRef]
11. Kihslinger, R.L.; Lema, S.C.; Nevitt, G.A. Environmental rearing conditions produce forebrain differences in wild Chinook salmon *Oncorhynchus tshawytscha*. *Comp. Biochem. Physiol. A* **2006**, *145*, 145–151. [CrossRef] [PubMed]
12. Spence, R.; Magurran, A.E.; Smith, C. Spatial cognition in zebrafish: The role of strain and rearing environment. *Anim. Cogn.* **2011**, *14*, 607–612. [CrossRef] [PubMed]
13. Papoutsoglou, S.E.; Mylonakis, G.; Miliou, H.; Karakatsouli, N.P.; Chadio, S. Effects of background colour on growth performances and physiological responses of scaled carp (*Cyprinus carpio* L.) reared in a closed circulated system. *Aquac. Eng.* **2000**, *22*, 309–318. [CrossRef]
14. Papoutsoglou, S.E.; Karakatsouli, N.; Louizos, E.; Chadio, S.; Kalogiannis, D.; Dalla, C.; Polissidis, A.; Papadopoulou-Daifoti, Z. Effect of Mozart's music (Romanze-Andante of "Eine Kleine Nacht Musik", sol major, K525) stimulus on common carp (*Cyprinus carpio* L.) physiology under different light conditions. *Aquac. Eng.* **2007**, *36*, 61–72. [CrossRef]

15. Martinez-Cardenas, L.; Purser, G.J. Effect of tank colour on Artemia ingestion, growth and survival in cultured early juvenile pot-bellied seahorses (*Hippocampus abdominalis*). *Aquaculture* **2007**, *264*, 92–100. [CrossRef]

16. Mesquita, F.O.; Godinho, H.P.; Azevedo, P.G.; Young, R.J. A preliminary study into the effectiveness of stroboscopic light as an aversive stimulus for fish. *Appl. Anim. Behav. Sci.* **2008**, *111*, 402–407. [CrossRef]

17. Monk, J.; Puvanendran, V.; Brown, J.A. Does different tank bottom colour affect the growth, survival and foraging behaviour of Atlantic cod (*Gadus morhua*) larvae? *Aquaculture* **2008**, *277*, 197–202. [CrossRef]

18. Cobcroft, J.M.; Battaglene, S.C. Jaw malformation in striped trumpeter *Latris lineate* larvae linked to walling behaviour and tank colour. *Aquaculture* **2009**, *289*, 274–282. [CrossRef]

19. El-Sayed, A.F.M.; El-Ghobashy, A.E. Effects of tank colour and feed colour on growth and feed utilization of thinlip mullet (*Liza ramada*) larvae. *Aquac. Res.* **2011**, *42*, 1163–1169. [CrossRef]

20. Basquill, S.P.; Grant, J.W.A. An increase in habitat complexity reduces aggression and monopolization of food by zebra fish (*Danio rerio*). *Can. J. Zool.* **1998**, *76*, 770–772. [CrossRef]

21. Carfagnini, A.G.; Rodd, F.H.; Jeffers, K.B.; Bruce, A.E.E. The effects of habitat complexity on aggression and fecundity in zebrafish (*Danio rerio*). *Environ. Biol. Fishes* **2009**, *86*, 403–409. [CrossRef]

22. Hasegawa, K.; Maekawa, K. Potential of habitat complexity for mitigating interference competition between native and non-native salmonid species. *Can. J. Zool.* **2008**, *86*, 386–393. [CrossRef]

23. Roberts, L.J.; Taylor, J.; Garcia de Leaniz, C. Environmental enrichment reduces maladaptive risk-taking behaviour in salmon reared for conservation. *Biol. Conserv.* **2011**, *144*, 1972–1979. [CrossRef]

24. Finstad, A.G.; Einum, S.; Forseth, T.; Ugedal, O. Shelter availability affects behaviour, size-dependent and mean growth of juvenile Atlantic salmon. *Freshwat. Biol.* **2007**, *52*, 1710–1718. [CrossRef]

25. Reed, B.; Jennings, M. *Guidance on the housing and care of zebrafish, Danio rerio*; RSPCA: Horsham, UK, 2010.

26. Kistler, C.; Hegglin, D.; Würbel, H.; König, B. Preference for structured environment on zebrafish (*Danio rerio*) and checker barbs (*Puntius oligolepis*). *Appl. Anim. Behav. Sci.* **2011**, *135*, 318–327. [CrossRef]

27. Corkum, L.D.; Cronin, D.J. Habitat complexity reduces aggression and enhances consumption in crayfish. *J. Ethol.* **2004**, *22*, 23–27. [CrossRef]

28. Hamilton, I.M.; Dill, L.M. Monopolization of food by zebrafish (*Danio rerio*) increases in risky habitats. *Can. J. Zool.* **2002**, *80*, 2164–2169. [CrossRef]

29. Larson, E.T.; O'Malley, D.M.; Melloni, R.H. Aggression and vasotocin are associated with dominant-subordinate relationships in zebrafish. *Behav. Brain Res.* **2006**, *167*, 94–102. [CrossRef]

30. Spence, R.; Gerlach, G.; Lawrence, C.; Smith, C. The behaviour and ecology of the zebrafish, *Danio rerio*. *Biol. Rev.* **2008**, *83*, 13–34. [CrossRef]

31. Watt, P.J.; Skinner, A.; Hale, M.; Nakagawa, S.; Burke, T. Small subordinate male advantage in the zebrafish. *Ethology* **2011**, *117*, 1003–1008. [CrossRef]

32. Spence, R.; Smith, C. Male territoriality mediates density and sex ratio effects on oviposition in the zebrafish, *Danio rerio*. *Anim. Behav.* **2005**, *69*, 1317–1323. [CrossRef]

33. Grant, J.W.A.; Kramer, D.L. Temporal clumping of food arrival reduces its monopolization and defense by zebrafish, *Brachydanio rerio*. *Anim. Behav.* **1992**, *44*, 101–110. [CrossRef]

34. McCarthy, I.D.; Carter, C.G.; Houlihan, D.F. The effect of feeding hierarchy on individual variability in daily feeding of rainbow trout, *Oncorhynchus mykiss*. *J. Fish Biol.* **1992**, *41*, 257–263. [CrossRef]

35. Johnson, D.W. Combined effects of condition and density on post-settlement survival and growth of a marine fish. *Oecologia* **2008**, *155*, 43–52. [CrossRef] [PubMed]

36. Samhouri, J.F. Food supply influences offspring provisioning but not density-dependent fecundity in a marine fish. *Ecology* **2009**, *90*, 3478–3488. [CrossRef]

37. Pickering, A.D.; Pottinger, T.G. Stress responses and disease resistance in salmonid fish: Effects of chronic elevation of plasma cortisol. *Fish Physiol. Biochem.* **1989**, *7*, 253–258. [CrossRef]

38. Fox, H.E.; White, S.A.; Kao, M.H.F.; Fernald, R.D. Stress and dominance in a social fish. *J. Neurosci.* **1997**, *17*, 6463–6469. [CrossRef]

39. Pottinger, T.G.; Pickering, A.D. The influence of social interaction on the acclimation of rainbow trout, *Oncorhynchus mykiss*, to chronic stress. *J. Fish Biol.* **1992**, *41*, 435–447. [CrossRef]

40. Wilkes, L.; Owen, S.F.; Readman, G.D.; Sloman, K.A.; Wilson, R.W. Does structural enrichment for toxicology studies improve zebrafish welfare? *Appl. Anim. Behav. Sci.* **2012**, *139*, 143–150. [CrossRef]

41. Keck, V.A.; Edgerton, D.S.; Hajizadeh, S.; Swift, L.L.; Dupont, W.D.; Lawrence, C.; Boyd, K.L. Effects of complexity on pair-housed zebrafish. *J. Am. Assoc. Lab. Anim. Sci.* **2015**, *54*, 378–383.
42. Korte, S.M.; Olivier, B.; Koolhaas, J.M. A new animal welfare concept based on allostasis. *Phys. Behav.* **2007**, *92*, 422–428. [CrossRef]
43. Næsje, T.F.; Thorstad, E.B.; Forseth, T.; Aursand, M.; Saksgård, R.; Finstad, A.G. Lipid class content as an indicator of critical periods for survival in juvenile Atlantic salmon (*Salmo salar*). *Ecol. Freshw. Fish* **2006**, *15*, 572–577. [CrossRef]
44. Von Krogh, K.; Sørensen, C.; Nilsson, G.E.; Øverli, Ø. Forebrain cell proliferation, behavior, and physiology of zebrafish, *Danio rerio*, kept in enriched or barren environments. *Physiol. Behav.* **2010**, *101*, 32–39. [CrossRef] [PubMed]
45. Garduño-Paz, M.V.; Couderc, S.; Adams, C.E. Habitat complexity modulates phenotype expression through developmental plasticity in the threespine stickleback. *Biol. J. Linn. Soc.* **2010**, *100*, 407–413. [CrossRef]
46. Brydges, N.M.; Braithwaite, V.A. Does environmental enrichment affect the behaviour of fish commonly used in laboratory work? *Appl. Anim. Behav. Sci.* **2009**, *118*, 137–143. [CrossRef]
47. Burns, J.G.; Saravanan, A.; Rodd, F.H. Rearing environment affects the brain size of guppies: Lab-reared guppies have smaller brains than wild-caught guppies. *Ethology* **2009**, *115*, 122–133. [CrossRef]
48. Réale, D.; Reader, S.M.; Sol, D.; McDougall, P.T.; Dingemanse, N.J. Intergrating animal temperament within ecology and evolution. *Biol. Rev.* **2007**, *82*, 291–318. [CrossRef]
49. Ariyomo, T.O.; Watt, P.J. The effect of variation in boldness and aggressiveness on the reproductive success of zebrafish. *Anim. Behav.* **2012**, *83*, 41–46. [CrossRef]
50. Höglund, E.; Balm, P.H.M.; Winberg, S. Behavioural and neuroendocrine effects of environmental background colour and social interaction in Arctic charr (Salvelinus alpines). *J. Exp. Biol.* **2002**, *205*, 2535–2543.
51. Rosemberg, D.B.; Rico, E.P.; Mussulini, B.H.M.; Paiato, A.L.; Calcagnotto, M.E.; Bonan, C.D.; Dias, R.E.; Blaser, R.E.; Souza, D.O.; de Oliveira, D.L. Differences in spatio-temporal behavior of zebrafish in the open tank paradigm after a short-period confinement into dark and bright environments. *PLoS ONE* **2011**, *6*, e19317. [CrossRef]
52. Gerlai, R.; Lahav, M.; Guo, S.; Rosenthal, A. Drinks like a fish: Zebra fish (*Danio rerio*) as a behaviour genetic model to study alcohol effects. *Pharmacol. Biochem. Behav.* **2000**, *67*, 773–782. [CrossRef]
53. Moretz, J.A.; Martins, E.P.; Robison, B.D. The effects of early and adult social environment on zebrafish (*Danio rerio*) behavior. *Environ. Biol. Fish.* **2007**, *80*, 91–101. [CrossRef]
54. Walsh, R.N.; Cummins, R.A. The open-field test: A critical review. *Psychol. Bull.* **1976**, *83*, 482–504. [CrossRef]
55. Moretz, J.A.; Martins, E.P.; Robison, B.D. Behavioral syndromes and the evolution of correlated behaviour in zebrafish. *Behav. Ecol.* **2007**, *18*, 556–562. [CrossRef]
56. Burns, J.G. The validity of three tests of temperament in guppies, *Poecilia reticulata*. *J. Comp. Psychol.* **2008**, *122*, 344–356. [CrossRef] [PubMed]
57. Westerfield, M. *The Zebrafish Book. A Guide for the Laboratory Use of Zebrafish (Danio rerio)*, 4th ed.; Chapman and Hall: London, UK, 1995.
58. Brand, M.; Granato, M.; Nüsslein-Volhard, C. Keeping and raising zebrafish. In *Zebrafish: A Practical Approach*; Nüsslein-Volhard, C., Dahm, R., Eds.; Oxford University Press: Oxford, UK, 2002; pp. 7–37.
59. Spence, R.; Ashton, R.; Smith, C. Oviposition decisions are mediated by spawning site quality in wild and domesticated zebrafish, *Danio rerio*. *Behaviour* **2007**, *144*, 953–966. [CrossRef]
60. Mertens, J. Year-round controlled mass reproduction of the zebrafish *Brachydanio rerio* (Hamilton-Buchanan). *Aquaculture* **1973**, *2*, 245–249. [CrossRef]
61. Gellert, G.; Heinrichsdorff, J. Effect of age on the susceptibility of zebrafish eggs to industrial waste water. *Water Res.* **2001**, *35*, 3754–3757. [CrossRef]
62. R Development Core Team. *R: A Language and Environment for Statistical Computing*; R Foundation for Statistical Computing: Vienna, Austria, 2011.

*fishes*

*Article*

# Welfare Challenges Influence the Complexity of Movement: Fractal Analysis of Behaviour in Zebrafish

**Anthony G. Deakin [1,2], Joseph W. Spencer [1], Andrew R. Cossins [2], Iain S. Young [2] and Lynne U. Sneddon [2,*]**

[1] Department of Electrical Engineering and Electronics, University of Liverpool, Liverpool L69 3GJ, UK; anthonyd@liverpool.ac.uk (A.G.D.); joe@liverpool.ac.uk (J.W.S.)
[2] Institute of Integrative Biology, University of Liverpool, Liverpool L69 7ZB, UK; cossins@liverpool.ac.uk (A.R.C.); isyoung@liverpool.ac.uk (I.S.Y.)
[*] Correspondence: lsneddon@liverpool.ac.uk; Tel.: +44-151-795-4388

Received: 19 December 2018; Accepted: 1 February 2019; Published: 7 February 2019

check for
updates

**Abstract:** The ability to assess welfare is an important refinement that will ensure the good condition of animals used in experimentation. The present study investigated the impact of invasive procedures on the patterns of movement of zebrafish (*Danio rerio*). Recordings were made before and after fin clipping, PIT tagging and a standard pain test and these were compared with control and sham handled zebrafish. The fractal dimension (FD) from the 3D trajectories was calculated to determine the effect of these treatments on the complexity of movement patterns. While the FD of zebrafish trajectories did not differ over time in either the control or sham group, the FDs of the treatment groups reduced in complexity. The FD of fish injected with different strengths of acetic acid declined in a dose-dependent manner allowing us to develop an arbitrary scale of severity of the treatments. The 3D trajectory plots from some groups indicated the presence of repetitive swimming patterns akin to stereotypical movements. When administered with lidocaine, which has analgesic properties, the movement complexity of fin clipped fish reverted to a pattern that resembled that of control fish. Fractal analysis of zebrafish locomotion could potentially be adopted as a tool for fish welfare assessment.

**Keywords:** *Danio rerio*; fractal analysis; nociception; pain; stereotypical behaviour

---

## 1. Introduction

The ability to monitor the welfare of experimental animals is a crucial refinement that can inform scientists of the severity of procedures. Here, we define animal welfare as the state of the individual as it attempts to cope with the environment [1,2]. However, determining an animal's welfare state is extremely difficult because animals cannot verbalise their internal experiences to experimenters. Instead scientists have to use indirect measures such as changes in behaviour and physiology to assess whether an animal's welfare has been compromised [3–6]. While there has been considerable work on assessing markers of rodent welfare [7,8] there has been relatively less research on fish. Currently fish are one of the most popular species used in experimentation, second only to mice, with approximately half a million individuals used in UK experiments in 2017 [9]. Zebrafish (*Danio rerio*) have become an important experimental model organism owing to the detailed genetic information available, the high genetic homology of zebrafish to humans, their relatively low cost, short generation time and easy maintenance and handling amongst other attributes [10–12]. Despite their popularity, the assessment and monitoring of zebrafish welfare during invasive experimentation is far behind that of rodents [13], thus, finding a means of accurately gauging the health status of fish would enhance the wellbeing of a large number of experimental animals.

Although the capacity of fish to perceive pain has been questioned [14,15] there is a steady accumulation of experimental evidence that fish meet the criteria for animal pain [5]. Fish possess nociceptors, the receptors required to detect pain-causing stimuli [16–19] as well as altered activity in brain areas that are activated at a molecular, physiological and functional level in response to pain [20–22]. Fish also show prolonged adverse changes in their behaviour and physiology in response to a noxious stimulus [23–27] that are ameliorated by providing analgesia [28–31]. For example, zebrafish injected subcutaneously in the lips with a noxious substance (acetic acid) showed a concomitant decrease in activity and dramatic increase in ventilation rate [25]. This evidence highlights the importance of detecting and characterising detrimental changes in the welfare of fish used in research, especially when commonly used procedures such as fin clip result in the damage of tissue containing nociceptors [26]. In order to characterise welfare changes from behaviour, a method for quantifying and comparing complex behavioural patterns in a broad, high-level but compressed manner is required.

Since animals continually evaluate their constantly changing environments and re-adjust their priorities accordingly through real-time actions, their natural movement trajectories are complex and stochastic rather than deterministic since three dimensions are available for flying/swimming. Conventional methods, for example Euclidian geometry, may describe the complex trajectories with high accuracy but in a highly complex manner [32–34]. If a reasonably accurate specification of a trajectory is required, machine learning methods may model or generate autonomous trajectories to a high degree of approximation, for example evolutionary computation [35], phase transition networks [36], Markov models [37] and support vector machines [38] among others. However, these methods retain some complexity and are therefore not usually easily directly deployable as comparative measures of complexity. Fractal dimension (FD) [39], Principal Component Analysis (PCA) [40] and Chromatic Analysis [41] are some of the techniques that assist with substantially reducing the dimensionality (complexity) of models and data. In the case of PCA and Chromatic Analysis, generally two, three or more parameters are identified to explain the main features of the behaviour. FD abstracts a single dimension from a data pattern in order to characterise its complexity in a single parameter, which may be used as a coarse, highly compressed measure of complexity for comparative purposes, at the expense of broad over-simplification of some of the detail of rich and diverse patterns that machine learning methods can capture. It is thereby considered to be a potentially suitable indicator for these purposes.

Fractal analysis can reveal patterns previously undetected by standard analysis of frequency/duration. It was originally developed to analyse geometric complexity using the concept of scaling [39] but has since been applied to the measurement of the temporal and spatial complexity of biological structures or systems [42,43]. Medicine in particular has benefited from using this approach, extracting 'hidden information' from physiological time series data, identifying for example, changing complexities in heart rate as a response to age and disease in humans where standard analysis revealed little detail [42]. Fractal analysis has also been applied in animal welfare studies to demonstrate a reduction in behavioural complexity linked to parasitic infection and pregnancy in Spanish ibex (*Capra pyrenaica*) [44], impaired health, ageing and low dominance status in Japanese macaques (*Macaca fuscata yakui*) [45], sickness in chimpanzees (*Pan troglodytes schweinfurthii*) [46,47], exposure to toxicants in fish and shrimps [48–50], welfare of marine diving mammals [51] and stress in dolphins (*Tursiops aduncus*) [52]. Surprisingly in most of these studies, standard behavioural approaches revealed little difference between treatments.

The aim of this study was to determine whether fractal analysis could reveal changes in female zebrafish behaviour in response to a series of potentially painful procedures. Different modes of pain elicit differing behavioural responses. For example, zebrafish administered with a potentially painful stimulus to the lip area did not show tail wafting which is observed during the same stimulus administered near to the tail fin (caudal peduncle) or after fin clipping [25,29,53]. Therefore, it is vital to assess more than one mode of pain in the present study. Previous work in mammalian species has highlighted a reduction in behavioural complexity as a result of stress/impaired health with speculation that behavioural complexity might correlate with states of increased allostatic load [46,48,52] or with pain and pre-pathological stress states [43,54,55]. The utility of fractal analysis to characterise the behavioural response of zebrafish to stress and to potentially painful stimuli could provide a useful tool for welfare assessment of laboratory fish. We hypothesized that potentially painful interventions in female zebrafish will also result in a reduction in behavioural complexity and that administering a pain-relieving drug would ameliorate these effects.

## 2. Results

There was a significant interaction between treatment group and time ($F_{28,192} = 2.757$, $p < 0.05$; Figure 1) indicating that the type of treatment had a differential effect on the resulting zebrafish fractal dimension over time. The effect of both treatment and time on zebrafish FD was then investigated as discussed below.

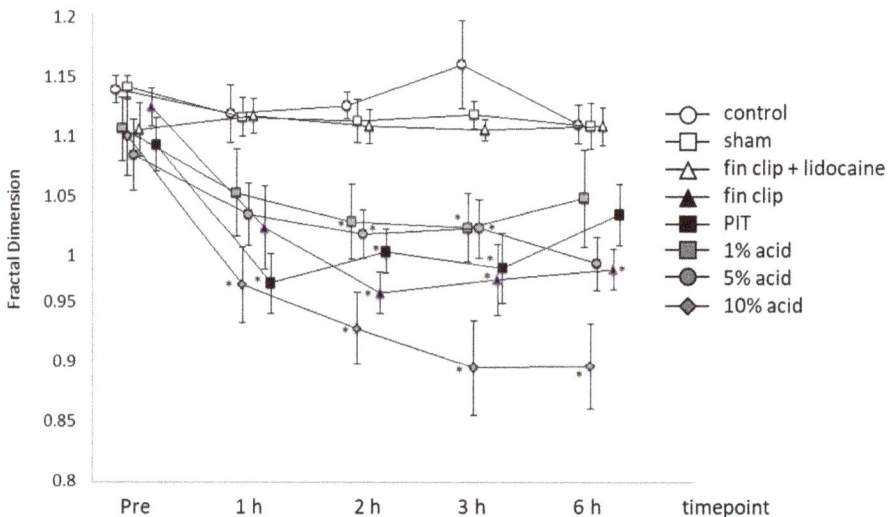

**Figure 1.** Mean (±S.E.) fractal dimension ($y$-axis) of zebrafish trajectories under the following treatments: control, sham handled, fin clip + lidocaine, fin clip, PIT tag, 1% acid lip, 5% acid lip and 10% acid lip ($n = 7$ each group) over time. * Significant difference ($p < 0.05$) when comparing treatment groups with controls. Data points have been altered via jitter to reduce overlap.

### 2.1. Effect of Treatment

Initially all fish had similar FD properties pre-treatment with no significant differences between any of the treatment groups ($F_{7,48} = 0.794$, $p = 0.596$; Figure 1) demonstrating lidocaine did not affect FD prior to fin clipping. However, after treatment there were changes in complexity at the 1 h ($F_{7,48} = 4.388$; $p < 0.05$; Figure 1), 2 h ($F_{7,48} = 9.780$; $p \leq 0.001$; Figure 1), 3 h ($F_{7,48} = 8.091$, $p \leq 0.001$; Figure 1) and 6 h ($F_{7,48} = 7.145$, $p < 0.05$; Figure 1) time points. Post hoc analysis revealed that control fish had more complex trajectories than the fin clip group at 2, 3 and 6 h ($p < 0.05$), the PIT tag group at 1, 2 and 3 h ($p < 0.05$), the 1% acid at 3h ($p < 0.05$), the 5% acid at 2 and 3 h ($p < 0.05$) and the 10% acid group at 1, 2, 3 and 6 h ($p < 0.05$). The effect of stress in the sham handled group did not result in FD that differed from controls but did result in much higher FD values than the PIT group at 1 and 2 h ($p < 0.05$), the fin clip group at 2, 3 and 6 h ($p < 0.05$), the 5% acid group at 2 h ($p < 0.05$) and the 10% acid group at 1, 2, 3 and 6 h ($p < 0.05$). The administration of lidocaine appeared to ameliorate the fin clip as the FD values in this group did not differ from controls ($p > 0.05$) but did differ from the fin clip group at 2 and 6 h ($p < 0.05$), the PIT tag group at 1 and 2 h ($p < 0.05$) and the 10% acid group at 1, 2, 3 and 6 h ($p < 0.05$). A visual analysis of 3D trajectory plots generated by treatment groups with high FD values (control fish) versus low FD values (Fin clip, PIT tag etc.) appeared distinctly different with 3D plots from fish with lower FD values indicating repetitive movement patterns (Figure 2).

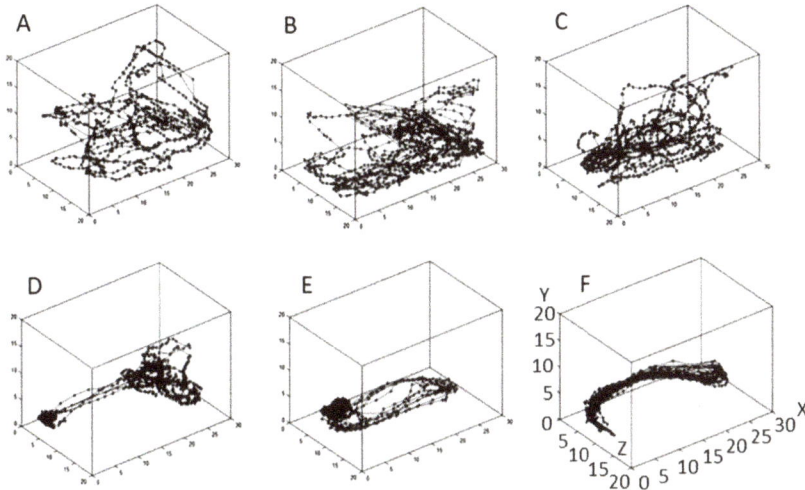

**Figure 2.** Comparison between the 3D trajectory plots (0–5 min sample taken from a 25 min recording at the 2 h time point) of three control zebrafish with fractal dimension (FD) scores above 1.15 (**A**, FD = 1.17; **B**, FD = 1.16; **C**, FD = 1.18) and three treatment zebrafish with FD below 0.9 (**D**, FD = 0.88 10% acid lip; **E**, FD = 0.90 fin clip; **F**, FD = 0.90 PIT tag). Those selected reflect that observed across each group. *x*, *y*, *z*-axes shown on plot F.

## 2.2. Effect of Time

Over the duration of the experiment there were no significant differences in the FDs of control ($F_{4,24} = 0.906$, $p = 0.476$), sham handled ($F_{4,24} = 1.294$, $p = 0.300$), Analgesic fin clipped fish treated with lidocaine ($F_{4,24} = 0.119$, $p = 0.975$) and 1% acid lip ($F_{4,24} = 1.158$, $p = 0.354$), therefore, this measure remained the same for the duration of the experiment. However, the FD of the fin clip ($F_{4,24} = 10.580$, $p \leq 0.0001$) PIT tag ($F_{4,24} = 6.112$, $p \leq 0.005$), 5% acid ($F_{4,24} = 3.149$, $p < 0.032$) and 10% acid group ($F_{4,24} = 7.967$, $p < 0.0001$) declined over the 1–3 h time points with some groups showing a slight increase in score at 6 h (except the 5 and 10% acetic acid groups). Pre intervention FD scores were significantly higher than all other post intervention time points including 6 h ($p < 0.05$).

## 2.3. Hypothetical Scale of Severity

The range of values associated with each treatment appeared to occupy a sliding scale with control (1.08–1.15; Figure 3) and 10% acid lip (0.83–0.97; Figure 3) occupying opposite ends of the spectrum while the rest of the treatments (Lidocaine 1.06–1.12; Sham handled 1.04–1.11; 1% Acid 0.96–1.09; PIT tag 0.94–1.05; 5% Acid 1.03–0.95 and Fin clip 1.02–0.9; Figure 3) overlapped each other across the range of the spectrum. The clustering of FD scores for each treatment combined with the observed decrease in FD as the percentage of acetic acid increased led to the tentative development of a potential scale of severity ranging from Normal (Control); Stress (Sham); to Mild to Severe Pain linked to FD scores from the painfully treated groups (Figure 3).

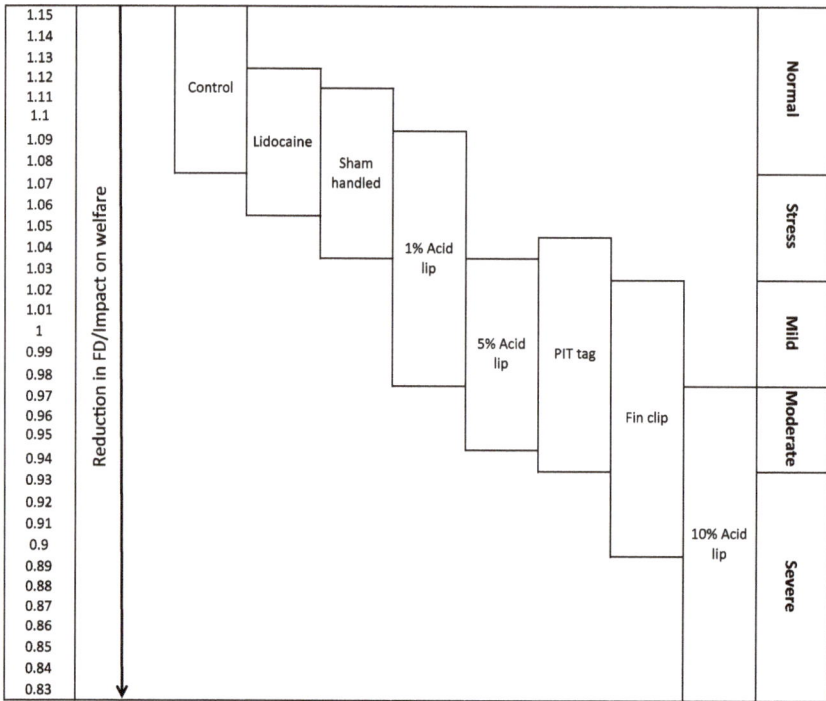

**Figure 3.** Hypothetical FD welfare scale indicating the range of FD values associated with each treatment group. A decrease in FD value indicates a reduction in welfare and an arbitrary scale of intensity represents zebrafish welfare as normal, stressed and in pain from mild to severe.

## 3. Discussion

Fractal analysis of the locomotory behaviour of female laboratory zebrafish subject to two laboratory procedures and a validated pain test was found to differentiate undisturbed fish from those subject to potentially painful procedures. The reduction in the complexity of the fish trajectories observed in this study signified a profound change in behaviour after a noxious event. While the trajectory complexity of control and sham handled fish remained relatively constant over the duration of the experiment, a significant reduction in complexity was noted across the pain groups. The decrease in movement complexity associated with the potentially painful interventions used in this study is similar to that observed in mammalian studies where acute stressors resulted in significant reductions in behavioural complexity [43,48,56]. This is also consistent in studies on pain in rainbow trout [23,28] and zebrafish [25,29,31,53] where the amount of swimming activity was reduced. Relatively lower behavioural complexity has been suggested to be an indicator of stereotypical behaviour [57]. Stereotypies are repetitive behaviours with no obvious function but may be linked to a sign that the animal is trying to cope with the current welfare challenging situation [58] or is in a state of distress [59]. In the present study the 3D trajectory plots of individuals with the greatest reduction in complexity showed repetitive swimming patterns. The presence of stereotypical behaviours in other species may indicate poor welfare [60] which can result from the application of a noxious stimuli [61].

An important property of a system used to identify changes in animal welfare is its ability to detect changes in behaviour that are proportional to the intensity of the applied stimulus. In this study the FD value of female zebrafish trajectories declined at a rate that scaled with the concentration of acetic acid. A previous study conducted in our laboratory [62] explored the effect of these same noxious interventions on more traditional behavioural measures including average speed (cm/s), percentage time spent in the bottom of the tank and amount of the tank explored. In that study the behavioural response of the 1% acid group was easily distinguishable from the two higher strengths as the higher strength groups had slower average speeds and spent more time on the bottom of the tank with less exploration: the difference between the 5 and 10% acid group, however, was much harder to discern but can be seen using FD analysis here. The only behavioural difference between the two higher strengths of acid was in the amount of time individuals spent in the top half of the tank; zebrafish injected with 10% acid spent more time in the top half of the tank rather than the bottom which was the case for all other treatment groups. This behavioural measure was similar to that observed in controls but inconsistent with that observed in the other pain groups (fin clip, PIT tag insertion, 1 and 5% acid groups) where avoiding the top half of the tank, an anti-predatory behaviour or anxiety response [63], seemed fairly predictable. This failure to avoid the top half of the tank by 10% Acid fish could be due to the severity of the pain being so great that it took priority over normal anti predatory or anxiety behaviour; this phenomenon has been seen in another fish species, the rainbow trout, where subcutaneous acid injection into the lips resulted in these fish not exhibiting normal fear or anti-predator responses [23,64]. The use of standard behaviour measures, therefore, required at least two behaviours (time spent in the bottom half of the tank and either average speed or percentage of tank explored) to identify a meaningful difference between the 5 and 10% acid groups [62] while here fractal analysis was able to differentiate the two groups with just one value. Current definitions of animal pain suggest animals may respond by altering their behaviour and when animals reduce activity this may be an attempt to conserve energy to divert to the healing process but may also be similar to guarding behaviour where reduced use of a limb or area prevents further damage and pain [5]. Reduced activity is observed in mammalian responses to pain and thus the data presented here and in other studies using trout and zebrafish [23,30,31,62] suggests this response is evolutionarily conserved and underlies mechanisms to recover from injury. In the laboratory context prolonged changes in behaviour could potentially confound experimental studies [65] whereas in a natural setting abnormal behaviours may alter the risk of being detected by predators [23].

The scaled decrease in movement complexity in relation to different strengths of acetic acid allowed the development of a provisional scale of severity that we then used to attempt to gauge the intensity of other commonly practiced procedures. Both the fin clip and PIT tag resulted in a reduction in complexity that fell between that observed for the 5 and 10% acid groups suggesting that these procedures are significantly painful. The changes in behaviour associated with these procedures, however, could also be due to the physical changes of the body that accompany these interventions. The impact of administering lidocaine which acts as a local anaesthetic with analgesic properties makes this explanation unlikely since fin clipped fish behaviour returned to normal when lidocaine was subsequently applied; their FD was not significantly different from control or sham handled zebrafish and did not differ over time. It is worth considering that lidocaine could have impacted normal behaviour although adverse behavioural reactions to this drug were not observed during the pre-treatment time point. Indeed, a study using lidocaine as an immersive agent in zebrafish demonstrated there were no side-effects when administering lidocaine alone [29].

The addition of extra weight via the implantation of a PIT tag appeared to have the least impact on swimming performance with several studies reporting no negative effects across a range of species on critical swimming velocity [65–68]; a measure of the swimming velocity at which maximum oxygen uptake occurs [69]. Although similar studies have not been conducted in zebrafish, the PIT tags used in this study equated to approximately 2% of the individual's bodyweight. This is much less than the tags (6–12% bodyweight) used in a study on chinook salmon (*Oncorhynchus tshawytscha*) which found no detrimental effects on swimming performance [66]. Measurements taken within our laboratory also suggest that female zebrafish can routinely carry eggs equalling 5+% of their bodyweight making it unlikely that PIT tag insertion in this study affected swimming performance. The reduction in behavioural complexity over the 6 h period, therefore, is likely to result from the tissue damage that accompanies this procedure. The potential for this procedure to be painful is also validated by another study from within our laboratory where the standard behavioural measurements (average speed, percentage tank explored and percentage time spent in the bottom half of the tank) of PIT tagged fish diverted from control behaviour to a similar degree to that observed in the 5% acid group across the 6 h experiment [62]. Previous studies have noted that in some species tagging is associated with infections, reduced feeding and increased mortality [67,68]. The immediate physiological response to the implantation of acoustic tags in carp and roach (*Rutilus rutilus*) led to an increase in cortisol concentrations with peak cortisol levels occurring between 2 and 10 h post-tagging [70]. Although the response to tagging appears to be species and tag specific [68] several studies have indicated the potential for tag implantation to induce an acute stress response and lead to mortality although no mortality was seen in the present study; the reduction in movement complexity noted in this study would certainly suggest that the procedure is having a detrimental effect on behaviour. While little has been done to address the potential for this procedure to cause pain, it is possible that the implantation of PIT tags in the short term results in damage to the musculature of the body wall leading to abdominal discomfort.

Tail or caudal fin clipping does have the potential to impede swimming ability when tested in other species [71,72]. Comparisons between wild type and no-tail zebrafish indicated the absence of a caudal fin could result in a 65% reduction in critical swimming performance [73]. This is an important consideration as a reduction in the ability to swim efficiently could be responsible for changes in more traditional behavioural measures such as locomotor activity, rather than being due to behaviours associated with an animal undergoing a painful treatment. One of the benefits of measuring movement or FD complexity, however, is that it is largely independent of behaviours related to swimming performance and instead describes how the fish chooses to explore the area [43,74] thereby possibly offering an insight into the subjective experience of the individual. Even if these procedures reduced critical swimming velocity, this should not dictate an effect on complexity. To control for the effect of any physical modifications on movement complexity we included a treatment group providing fin clipped fish with lidocaine. The lidocaine appeared to ameliorate the behavioural effects of the fin clip in that movement complexity was restored to those seen in control fish. The analgesic properties of lidocaine have previously been demonstrated in trout, where the behavioural and physiological symptoms associated with the injection of acetic acid were greatly reduced [28] and in fin clipped zebrafish [29]. The ability of lidocaine to diminish the impact of the fin clip in this study is evidence that the reduction in behavioural complexity was not due to the physical change accompanying the removal of the fin but instead part of a complex behavioural reaction to painful tissue damage. This was mirrored in our previous studies where the fin clip elicited reductions in average speed, tank exploration and time spent in the bottom half of the tank [29,62]; these changes in behaviour were greatly ameliorated via the addition of lidocaine and other analgesics highlighting both the success of lidocaine as an analgesic as well as the ability of FD analysis to accurately describe the welfare status of zebrafish.

The potential for the fin clip to be painful is considerable as studies have demonstrated the presence of nociceptors within the tail fin of the common carp (*Cyprinus carpio*) [26]. Nociceptors are sensory neurons that are preferentially stimulated by noxious stimuli. The properties of nociceptors in fish have been well documented with their physiology being identical to that of mammalian nociceptors [18,75]. The nociceptors found in the tail fin of carp consisted of C-fibres and A-δ fibres at a similar abundance to those found in the trigeminal nerve in trout [26,75]. The removal of a considerable section of tail fin is therefore likely to excite these receptors while exposing them to the environment. As well as the identification of nociceptors, Roques et al. [26] also observed a distinct change in the behaviour and physiology of fin clipped Nile tilapia (*Oreochromis niloticus*), with individuals showing higher swimming activity while spending more time in the light than controls over the 6 h duration of the experiment. The fin clip procedure in the present study also caused a distinct change in behaviour relative to control and sham handled zebrafish with a reduction in movement complexity also lasting for the 6 h observation period.

Zebrafish are a gregarious species and are often held in groups. Studies have demonstrated that group housed zebrafish recover more quickly from fin clipping than those held individually [76] and that fear responses are reduced when zebrafish have olfactory and more importantly visual cues of conspecifics [77,78]. This phenomenon is termed social buffering where social support assists in reducing responses to threatening stimuli or events and appears evolutionarily conserved from fish to mammals [77,78]. Thus, it is conceivable that if we repeated these experiments in group housed zebrafish we may see a lesser or no change in the complexity of movement. However, we chose to test individual zebrafish as this is more relevant to the laboratory context. Zebrafish are held individually after invasive procedures to allow recovery and promote healing (e.g., cardiac surgery [79], optic nerve crush [80] and spinal lesions [81]) or to allow genotyping from fin clips [29]. Further, the 3D tracking of groups of animals whilst maintaining individual identity is not currently possible as demonstrated in the present study for individual behavioural tracking. More importantly, it would be very difficult to understand the complexities of the group response if we did not first understand the changes in an individual so we propose this is a stepping stone for future studies which should investigate this phenomenon in groups of zebrafish when technology allows.

The differential response of female zebrafish to the treatments in this study, indicate the ability of some interventions to have a greater impact on behavioural complexity than others; fin clip and 10% acetic acid interventions for example, exhibited much lower FD values relative to controls over time compared to the sham handled or 1% acid lip groups. Identifying the range of values associated with each intervention in this study allowed the tentative construction of a hypothetical FD based welfare scale by which the severity of new interventions could be measured. In this present study there is a clear gap between the ranges of values associated with control fish (1.08–1.15) versus those associated with fin clip (1.02–0.9) and 10% acid lip (0.94–0.83). From this it could be argued that values close to and above 1.15 indicate excellent welfare and those close to and below 0.9 indicate negative welfare. Sham handled fish occupied a range of values from 1.04–1.11 potentially indicating a part of the scale that might describe the stress of handling and anaesthesia. The range of values associated with the 1% acid lip (0.96–1.09) group occupy the middle of the scale slightly encroaching on to the range of control, PIT and fin clipped fish indicating a more mild to moderate impact on zebrafish welfare. From this scale it could be argued that in terms of the severity of procedure, 10% Acid Lip > Fin clip > PIT > 5% Acid Lip > 1% Acid lip>Sham handled. Future work could refine and add clarity to the scale through the testing of a wider range of procedures and could provide a means of assessing actual severity of a variety of invasive experiments including non-painful treatments that may cause stress or distress. It would also be vital to extend this work to male zebrafish and to other strains where it is possible behavioural responses may differ from AB females.

## 4. Materials and Methods

### 4.1. Subjects and Husbandry

Eight month old female zebrafish (*D. rerio*) ($n$ = 56; mean size 0.83 g $\pm$ 0.04) of AB strain were randomly selected from the University of Liverpool aquarium in-house breeding project: Stock fish were maintained in a semi-closed recirculation system in 10 L tanks at 28 $\pm$ 1 °C, with constant aeration on a 14:10 h light: dark cycle. The use of females removed the confounding factor of sex. Fish were selected at random, netted carefully into a 3L tank and transferred individually to a semi-closed recirculation system consisting of two parallel rows of nine glass tanks (20 $\times$ 30 $\times$ 20 cm; $n$ = 1 fish per tank) each fitted with an identical, external laminated printout of a green plant background. The background acted as a green screen enabling the behavioural tracking system [82] to accurately differentiate between the focal fish and the background. All tanks were supplied with filtered water (pH 7.2, $NH_3 \leq 0.01$ mg/L, $NO_2 \leq 0.01$ mg/L, $NO_3 \leq 5$ mg/L) maintained at a constant temperature of 28 $\pm$ 1°C, under a 14:10 light: dark regime with aeration provided by an aerated, fluidised 200 L biological filter. Fish were acclimatised in their experimental tank for two weeks prior to experimentation and fed twice daily *ad libitum* with a commercial tropical ornamental flake (TetraMin, Tetra, Melle, Germany). Fish were only used in experiments if they fed readily when food was presented for at least seven days. Fish were in chemical (through shared water) and visual contact with adjacent tanks so they had social contact until the evening prior to experimentation when two opaque pieces of plastic were placed in between tanks to visually isolate the test individuals.

*4.2. Treatment Groups*

The effect of several potentially painful procedures on the behaviour of zebrafish was tested against control (undisturbed), sham handled (anaesthetised and handled in a similar manner and time frame but no treatment) plus a further group that had a fin clip but were administered with lidocaine (5 mg/L dissolved in the tank water) which prevented the behavioural changes associated with fin clipping in a previous study on zebrafish [29]. Fish were randomly assigned to one of the eight treatment groups (*n* = 7 for each group): Control; sham handled; five noxiously-stimulated groups (1–3. Injected subcutaneously with either 1, 5 or 10% acetic acid into the lips; 4. PIT tag injection through the abdomen; and 5. fin clip where 40% of the caudal fin was removed as described in The Zebrafish Handbook (http://zfin.org/zf_info/zfbook/chapt7/7.8.html).); and an Analgesic group subject to fin clip administered with lidocaine, a local anaesthetic with pain-relieving properties (5 mg/L Sigma-Aldrich Co., Dorset, UK). Only the fin clip group were tested with lidocaine to keep sample sizes to a minimum and previous studies have demonstrated drugs with analgesic properties prevent behavioural changes in response to acetic acid [27,30,31,53,83]. Control fish were left undisturbed for the duration of the experiment; all other treatment fish were carefully netted and transferred to a 1 L beaker containing 500 mL of aerated water dosed with benzocaine (0.033 g L$^{-1}$; Sigma-Aldrich Co., Dorset, UK) where fish were anaesthetised to deep plane anaesthesia so they were unconscious during the procedure. Benzocaine was used as it has short lasting analgesic properties [84]. After anaesthesia the sham treatment group were handled similarly but without any invasive treatment applied. During anaesthesia the Acid groups were injected subcutaneously into the frontal lips using 2 µL per lip with either 1, 5 or 10% acetic acid using a sterile gastight syringe and needle (34 g; Hamilton; Bonaduz, Switzerland). PIT tag treatment fish were orientated upside down and a sterile 20 gauge needle used to inject a 4 mm PIT tag into the abdomen (Loligo systems, Viborg, Denmark). All fish were returned to their home tank after the intervention and allowed to recover from the anaesthesia where video recordings began 1 h afterwards; no mortalities occurred in response to any of the above treatments.

*4.3. Data Collection*

Fish movements were captured on video for 25 min periods, five times throughout the experiment at the following time points; 40 min pre-treatment, 1, 2, 3 and 6 h after treatment. These time points were chosen as fish subject to noxious stimulation usually show an initial adverse response up until 3 h then recovery by six hours in acetic acid tests [23,25]. Fish were tracked using two industrial IDS USB 3.0 colour video cameras (IDS; Obersulm, Germany) fitted with a 25 mm monofocal lens and connected to a computer (HP compact elite 8300; Palo Alto, CA, USA) running tracking software developed at the University of Liverpool [82]. Cameras positioned dorsally and laterally to the focal tank were used to track the 3D trajectories of fish. Cameras positioned above the tanks were mounted on a sliding gantry 1.4 m above the two parallel rows of nine tanks; this enabled the cameras to be moved from tank to tank without disturbing the fish. Cameras positioned laterally were attached to tripods 1.4 m away from the focal tanks and were manually moved between tanks although only one tank on each side was recorded each day with cameras moved the previous evening. Treatments were randomised to prevent order effects. Data files generated by the 3D tracking software written in MATLAB were used to analyse the fractal dimension of fish trajectories. The tracking software was validated through blind comparisons with a human observer comparing the scores percentage (%) time spent in different zones of the tank and % time spent inactive (8 videos tested with 100% accuracy). The pre-treatment behavioural recordings were carried out at the same time each day (commencing at 10:00 a.m. GMT), to minimise any effect of diurnal fluctuations on behaviour. Inflow to all tanks was turned off at 9:30 a.m. GMT to isolate each tank chemically and in the Analgesia group the lidocaine was added at this time to allow uptake and to ensure lidocaine in itself did not affect pre-treatment behaviour.

*4.4. Fractal Dimension*

Fractal dimensions (FDs) were obtained for the fish in order to characterise, in a single parameter, various aspects of their behaviours during each 25 min period under study (see Supplementary Information for raw data). The basic procedure for obtaining an FD followed that of Nimkerdphol and Nakagawa [33,85]. An FD may be obtained from a variety of source data, for example location or distance. In this case, three FDs, FD $(x, y, z)$, were obtained from the location of a fish in the tank along each of the $x$, $y$ and $z$ dimensions and an average composite FD was calculated for the three dimensions FD$x$, FD$y$, FD$z$ taken together. The procedure for ascertaining the fractal dimension for a period consisted of obtaining discrete Fourier transforms of the $x$, $y$ and $z$ vectors computed with the fast Fourier transform (FFT) [86] using the first 17 min (8192 location samples) out of the 25 min recording period (the number of samples in the FFT needs to be a power of 2. With the sampling rate used, 17 min = 8192 samples). A Fourier transform is usefully applied to signals in order to obtain a different view of them which can give additional information. Often, the transform converts the signal from one domain (e.g., time) to another (e.g., frequency). For example, a sound signal may be graphically represented as the amplitudes of the sampled signal on the $y$-axis with the $x$-axis as the time domain. In the case of the FD in the present study, application of the FFT in effect transforms the domain from being the absolute location coordinate of the fish in time to the domain of relative change in location coordinate. Figure 4 shows an example of the $x$-coordinate dimensions and the FFTs thereof for a Control fish and a Fin Clip fish 2 h after treatment.

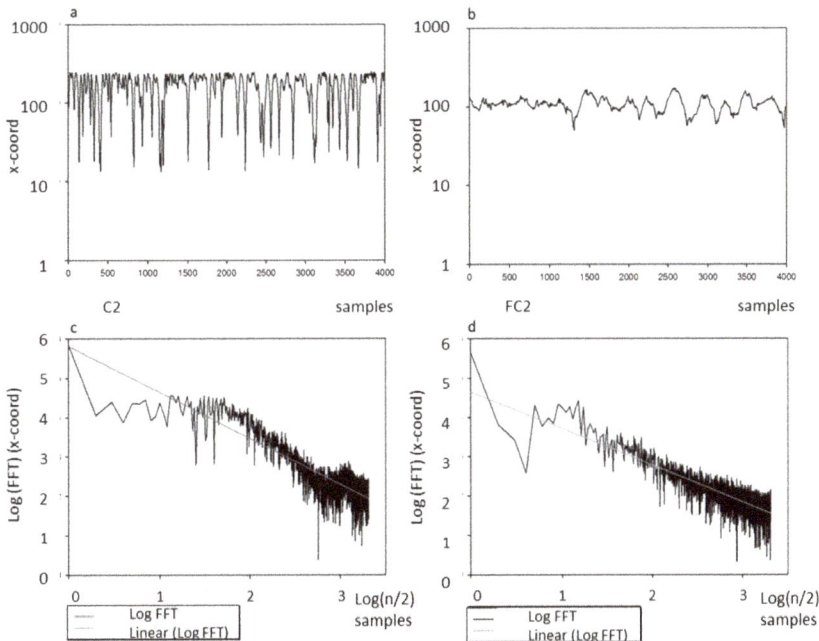

**Figure 4.** X-coordinates over 4096 samples (4000 shown) for a control zebrafish (**a**) and a fin clipped zebrafish (**b**) in period 3 together with the plots (**c**,**d**) of the logs of the FFTs of the $x$-coordinates with samples shown on the horizontal axis in the format log10(n/2) on plots (**c**,**d**).

Figure 4a,b show the *x*-coordinates of a control fish (C2) and a fin-clip fish (FC2) in their tank 2 h after intervention, indicating that C2 is much more active as compared with FC2 in both the extent of the trajectories across the tank within $x = 13$ to $x = 206$ and the frequency of them, whereby the maximum trajectories by FC2 are within $x = 49$ to $x = 171$ and are fewer in frequency. The gradient of the plot of an FFT of a variable (here the *x*-dimension) gives an indication of the complexity of the variable (here, activity in that dimension) and indicates the FD. For example, a static fish has a flat, constant trace across the samples (i.e., a straight line in Figure 4a or b and a corresponding FFT gradient (FD) of ~0. A fish that swims half way across the tank and back would have FD ~0.5, one that swims fully across, ~0.92 and one that swims two full round trips across and back, FD ~0.94. The FDs of C2 and FC2 in Figure 4c,d are 1.17 and 0.93 respectively, indicating that the complexity (FD) of the control fish activity by location is substantially higher than that of the fin clip fish.

*4.5. Statistical Analysis*

Data were analysed using SPSS software 21. The fractal dimension data was normally distributed for all interventions as assessed by Shapiro-Wilk's test ($p > 0.05$), displayed homogeneity of variance as assessed by Levene's test ($p > 0.05$) and did not violate sphericity ($p > 0.05$). An interaction between time and treatment was first tested on the FD via a mixed model ANOVA which also analysed the impact of treatment group and time separately followed by post hoc Tukey tests to determine where differences lay. A FD scale describing the welfare state of zebrafish was constructed using the range of FD values for each treatment group. The range of FD values for each treatment group was calculated using the means of the top and bottom third of all FD values post intervention. A hypothetical scale of severity was created based upon the FD calculations.

*4.6. Ethics Note*

Experiments were conducted with approval from the Home Office, U.K. (licence no. 40/3534) and the University of Liverpool's Ethics committee. At the end of the experiment, fish were euthanized using a schedule 1 method (concussion followed by pithing) and tissue harvested for use in other studies. All fish were treated humanely and care taken when carrying out the treatments. As different modes of pain elicit different behavioural responses it was vital we tested more than one type of painful stimulus. The treatments were chosen as they represent either a standard pain test, such as the acetic acid groups [23,27,31] or they are routinely used for identification purposes such as the fin clip [87,88] and PIT tagging [67]; thus knowledge of the severity of these procedures would be useful in refining procedures and applicable to other laboratories using these methods. The PIT tags used weighed 0.020 g equating to around 2% of the bodyweight of the individuals used in this study which is below the threshold weight of tags known to affect swimming performance [66]. Although there is evidence that benzocaine may be more aversive than other anaesthetics [89], benzocaine also acts as a local anaesthetic and so provides a period of short-term pain relief pre-operatively thus being the more ethical choice of anaesthetic during these painful treatments. The sham treated group controls for any stress associated with handling and anaesthesia. We did not perform any sham injections of non-painful saline since previous studies demonstrate there are no differences between control and sham injected fish [19,31,53,83].

## 5. Conclusions

The results from this study demonstrate the ability of fractal analysis to identify differences in behaviour that scale with the intensity of the administered noxious procedure thus highlighting its potential to reveal insights into the welfare status of zebrafish. This ability to differentiate between different strengths of acid appeared to give it a better resolution than the more traditional behavioural scores. Significant reductions in movement complexity were observed in most pain groups relative to controls and sham handled fish, which is consistent with what has been observed in mammals experiencing high levels of acute stress [43,44,52]. The greatest reductions in complexity were observed in the 10% acid lip and fin clip groups highlighting the possibility that these interventions may be more intense while the provision of lidocaine ameliorated the impact of the fin clip further highlighting the likelihood that these routine procedures are indeed painful. This data, therefore, demonstrates these procedures should be accompanied by the provision of analgesia and here we can recommend the use of 5 mg/L lidocaine based on the present study. Currently, procedures which result in acute pain for a few hours are deemed to be of mild severity under EU legislation [90] but our results show that the responses to fin clipping and PIT tagging persists for several hours and as such should be deemed moderately severe. However, if immersion analgesia is provided to alleviate any associated pain and discomfort, this procedure could be reduced to mild. The increased use of fish and in particular zebrafish in research means that they are subject to a wide variety of invasive procedures with little known about the ability of these interventions to cause pain. It is crucial, therefore, that non-invasive tools like fractal dimension analysis be used and developed to address the severity of protocols so that appropriate actions can be taken to ameliorate their impact upon health and welfare. Being able to assess severity and possible pain from a laboratory procedure and further minimising pain via analgesia represent important refinements in the treatment of experimental zebrafish.

**Supplementary Materials:** The following are available online at http://www.mdpi.com/2410-3888/4/1/8/s1. We supply the FD data set.

**Author Contributions:** L.U.S., A.R.C., J.W.S. and I.S.Y. obtained the funding; L.U.S., A.G.D. and A.R.C. designed the experiments; A.G.D. created the fractal analysis software; A.G.D. analysed the data with advice from L.U.S.; L.U.S. wrote the manuscript with assistance from A.G.D. and with approval of final draft by all authors.

**Funding:** We are grateful for research grant NC/K000888/1 from the National Centre for the Replacement, Refinement and Reduction of Animals in Research, UK (NC3Rs).

**Acknowledgments:** We thank Jonathan Buckley for assistance with the experiments and Samantha Moss and Rhys Sweeney for technical assistance.

**Conflicts of Interest:** The authors declare no competing or financial interests. The funders played no role in designing the experiments, their execution, data analysis, the writing of the paper or in the decision to publish.

## References

1. Broom, D.M. Indicators of poor welfare. *Brit. Vet. J.* **1986**, *142*, 524–526. [CrossRef]
2. Broom, D.M. Animal welfare: Concepts and measurement. *J. Anim. Sci.* **1991**, *69*, 4167–4175. [CrossRef] [PubMed]
3. Molony, V.; Kent, J.E.; McKendrick, I.J. Validation of a method for assessment of an acute pain in lambs. *Appl. Anim. Behav. Sci.* **2002**, *76*, 215–238. [CrossRef]
4. Sneddon, L.U. Pain in aquatic animals. *J. Exp. Biol.* **2015**, *218*, 967–976. [CrossRef] [PubMed]
5. Sneddon, L.U.; Elwood, R.W.; Adamo, S.A.; Leach, M.C. Defining and assessing animal pain. *Anim. Behav.* **2014**, *97*, 201–212. [CrossRef]
6. Zimmerman, M. Physiological mechanisms of pain and its treatment. *Klinische Anäesthesiologie Intensivtherapie* **1986**, *32*, 1–19.
7. Leach, M.C.; Klaus, K.; Miller, A.L.; di Perrotolo, M.S.; Sotocinal, S.G.; Flecknell, P.A. The Assessment of Post-Vasectomy Pain in Mice Using Behaviour and the Mouse Grimace Scale. *PLoS ONE* **2012**, *7*, e35656. [CrossRef] [PubMed]
8. Roughan, J.V.; Wright-Williams, S.L.; Flecknell, P.A. Automated analysis of postoperative behaviour: Assessment of HomeCageScan as a novel method to rapidly identify pain and analgesic effects in mice. *Lab. Anim.* **2009**, *43*, 17–26. [CrossRef] [PubMed]
9. Home Office. 2018. Available online: https://www.gov.uk/government/statistics/statistics-of-scientific-procedures-on-living-animals-great-britain-2017 (accessed on 18 December 2018).
10. Clark, K.J.; Ekker, S.C. How Zebrafish Genetics Informs Human Biology. *Nat. Educ.* **2015**, *8*, 3.
11. Hart, P.C.; Bergner, C.L.; Egan, R.J.; LaPorte, J.L.; Smolinsky, A.N.; Amri, H.; Zukowska, Z.; Glasgow, E.; Kalueff, A.V. The Utility of Zebrafish in Stress Research. *FASEB J.* **2009**, *23*, 1.
12. Hill, A.J.; Teraoka, H.; Heideman, W.; Peterson, R.E. Zebrafish as a model vertebrate for investigating chemical toxicity. *Toxicol. Sci.* **2005**, *86*, 6–19. [CrossRef] [PubMed]

13. Johansen, R.; Needham, J.R.; Colquhoun, D.J.; Poppe, T.T.; Smith, A.J. Guidelines for health and welfare monitoring of fish used in research. *Lab. Anim.* **2006**, *40*, 323–340. [CrossRef] [PubMed]

14. Iwama, G.K. The welfare of fish. *Dis. Aquat. Org.* **2007**, *75*, 155–158. [CrossRef] [PubMed]

15. Rose, J.D. The Neurobehavioral nature of fishes and the question of awareness and pain. *Rev. Fish. Sci.* **2002**, *10*, 1–38. [CrossRef]

16. Ashley, P.J.; Sneddon, L.U.; McCrohan, C.R. Properties of corneal receptors in a teleost fish. *Neurosci. Lett.* **2006**, *410*, 165–168. [CrossRef] [PubMed]

17. Ashley, P.J.; Sneddon, L.U.; McCrohan, C.R. Nociception in fish: Stimulus-response properties of receptors on the head of trout *Oncorhynchus mykiss*. *Brain Res.* **2007**, *1166*, 47–54. [CrossRef] [PubMed]

18. Sneddon, L.U. Trigeminal somatosensory innervation of the head of a teleost fish with particular reference to nociception. *Brain Res.* **2003**, *972*, 44–52. [CrossRef]

19. Sneddon, L.U.; Braithwaite, V.A.; Gentle, M.J. Do fishes have nociceptors? Evidence for the evolution of a vertebrate sensory system. *Proc. R. Soc. B-Biol. Sci.* **2003**, *270*, 1115–1121. [CrossRef] [PubMed]

20. Dunlop, R.; Laming, P. Mechanoreceptive and nociceptive responses in the central nervous system of goldfish (*Carassius auratus*) and trout (*Oncorhynchus mykiss*). *J. Pain* **2005**, *6*, 561–568. [CrossRef] [PubMed]

21. Reilly, S.C.; Kipar, A.; Hughes, D.J.; Quinn, J.P.; Cossins, A.R.; Sneddon, L.U. Investigation of Van Gogh-like 2 mRNA regulation and localisation in response to nociception in the brain of adult common carp (*Cyprinus carpio*). *Neurosci. Lett.* **2009**, *465*, 290–294. [CrossRef] [PubMed]

22. Reilly, S.C.; Quinn, J.P.; Cossins, A.R.; Sneddon, L.U. Novel candidate genes identified in the brain during nociception in common carp (*Cyprinus carpio*) and rainbow trout (*Oncorhynchus mykiss*). *Neurosci. Lett.* **2008**, *437*, 135–138. [CrossRef] [PubMed]

23. Ashley, P.J.; Ringrose, S.; Edwards, K.L.; Wallington, E.; McCrohan, C.R.; Sneddon, L.U. Effect of noxious stimulation upon antipredator responses and dominance status in rainbow trout. *Anim. Behav.* **2009**, *77*, 403–410. [CrossRef]

24. Millsopp, S.; Laming, P. Trade-offs between feeding and shock avoidance in goldfish (*Carassius auratus*). *Appl. Anim. Behav. Sci.* **2008**, *113*, 247–254. [CrossRef]

25. Reilly, S.C.; Quinn, J.P.; Cossins, A.R.; Sneddon, L.U. Behavioural analysis of a nociceptive event in fish: Comparisons between three species demonstrate. *Appl. Anim. Behav. Sci.* **2008**, *114*, 248–259. [CrossRef]

26. Roques, J.A.C.; Abbink, W.; Geurds, F.; van de Vis, H.; Flik, G. Tailfin clipping, a painful procedure Studies on Nile tilapia and common carp. *Physiol. Behav.* **2010**, *101*, 533–540. [CrossRef] [PubMed]

27. Sneddon, L.U. The evidence for pain in fish: The use of morphine as an analgesic. *Appl. Anim. Behav. Sci.* **2003**, *83*, 153–162. [CrossRef]

28. Mettam, J.J.; Oulton, L.J.; McCrohan, C.R.; Sneddon, L.U. The efficacy of three types of analgesic drugs in reducing pain in the rainbow trout, *Oncorhynchus mykiss*. *Appl. Anim. Behav. Sci.* **2011**, *133*, 265–274. [CrossRef]

29. Schroeder, P.; Sneddon, L.U. Exploring the efficacy of immersion analgesics in zebrafish using an integrative approach. *Appl. Anim. Behav. Sci.* **2017**, *187*, 93–102. [CrossRef]

30. Lopez-Luna, J.; Al-Jubouri, Q.; Al-Nuaimy, W.; Sneddon, L.U. Activity reduced by noxious chemical stimulation is ameliorated by immersion in analgesic drugs in zebrafish. *J. Exp. Biol.* **2017**, *220*, 1451–1458. [CrossRef] [PubMed]

31. Taylor, J.C.; Dewberry, L.S.; Totsch, S.K.; Yessick, L.R.; DeBerry, J.J.; Watts, S.A.; Sorge, R.E. A novel zebrafish-based model of nociception. *Physiol. Behav.* **2017**, *174*, 83–88. [CrossRef] [PubMed]

32. Bershadskii, A. An universal relation between fractal and Euclidean (topological) dimensions of random systems. *Eur. Phys. J. B* **1998**, *6*, 381–382. [CrossRef]

33. Nimkerdphol, K.; Nakagawa, M. Effect of sodium hypochlorite on zebrafish swimming behavior estimated by fractal dimension analysis. *J. Biosci. Bioeng.* **2008**, *105*, 486–492. [CrossRef] [PubMed]

34. Power, W.L.; Tullis, T.E. Euclidean and fractal models for the description of rock surface-roughness. *J. Geophys. Res.-Solid Earth Planets* **1991**, *96*, 415–424. [CrossRef]

35. Deakin, A.G.; Yates, D.F. Evolving and Optimizing Autonomous Agents' Strategies with Genetic Programming. In *Genetic Programming 1998: Proceedings of the Third Annual Conference, University of Wisconsin, Madison, WI, USA, 22–25 July 1998*; Koza, J.R., Banzhaf, W., Chellapilla, K., Deb, K., Dorigo, M., Fogel, D.B., Garzon, M.H., Goldberg, D.E., Iba, H., Riolo, R., Eds.; Morgan Kaufmann: San Francisco, CA, USA, 1998; pp. 42–47.

36. Deakin, A.G.; Yates, D.F. Phase Transition Networks: A Modelling Technique supporting the Evolution of Autonomous Agents' Tactical and Operational Activities. In *Lecture Notes in Computer Science, Proceedings of the AISB 97 Evolutionary Computing Workshop, University of Manchester, Manchester, UK, 7–8 April 1997*; Corne, D., Shapiro, J., Eds.; Springer Verlag: Berlin, Germany, 1997; Volume 1305, pp. 263–273.

37. Eddy, S.R. What is a hidden Markov model? *Nat. Biotechnol.* **2004**, *22*, 1315–1316. [CrossRef] [PubMed]

38. Cortes, C.; Vapnik, V. Support-vector networks. *Mach. Learn.* **1995**, *20*, 273–297. [CrossRef]

39. Mandelbrot, B. How Long Is the Coast of Britain? Statistical Self-Similarity and Fractional Dimension. *Science* **1967**, *156*, 636–638. [CrossRef] [PubMed]

40. Pearson, K. On lines and planes of closest fit to systems of points in space. *Philos. Mag.* **1901**, *2*, 559–572. [CrossRef]

41. Jones, G.R.; Deakin, A.G.; Spencer, J.W. *Chromatic Monitoring of Complex Conditions*; CRC Press-Taylor & Francis Group: Boca Raton, FL, USA, 2008.

42. Goldberger, A.L.; Amaral, L.A.N.; Hausdorff, J.M.; Ivanov, P.C.; Peng, C.K.; Stanley, H.E. Fractal dynamics in physiology: Alterations with disease and aging. *Proc. Nat. Acad. Sci. USA* **2002**, *99*, 2466–2472. [CrossRef] [PubMed]

43. Rutherford, K.M.D.; Haskell, M.J.; Glasbey, C.; Jones, R.B.; Lawrence, A.B. Fractal analysis of animal behaviour as an indicator of animal welfare. *Anim. Welf.* **2004**, *13*, S99–S103.

44. Alados, C.L.; Escos, J.M.; Emlen, J.M. Fractal structure of sequential behaviour patterns: An indicator of stress. *Anim. Behav.* **1996**, *51*, 437–443. [CrossRef]

45. MacIntosh, A.J.J.; Alados, C.L.; Huffman, M.A. Fractal analysis of behaviour in a wild primate: Behavioural complexity in health and disease. *J. R. Soc. Interface* **2011**, *8*, 1497–1509. [CrossRef] [PubMed]

46. Alados, C.L.; Huffman, M.A. Fractal long-range correlations in behavioural sequences of wild chimpanzees: A non-invasive analytical tool for the evaluation of health. *Ethology* **2000**, *106*, 105–116. [CrossRef]

47. Burgunder, J.; Pafco, B.; Petrzelkova, K.J.; Modry, D.; Hashimoto, C.; MacIntosh, A.J J Complexity in behavioural organization and strongylid infection among wild chimpanzees. *Anim. Behav.* **2017**, *129*, 257–268. [CrossRef]

48. Alados, C.L.; Weber, D.N. Lead effects on the predictability of reproductive behavior in fathead minnows (*Pimephales promelas*): A mathematical model. *Environ. Toxicol. Chem.* **1999**, *18*, 2392–2399. [CrossRef] [PubMed]

49. Eguiraun, H.; Lopez-de-Ipina, K.; Martinez, I. Application of Entropy and Fractal Dimension Analyses to the Pattern Recognition of Contaminated Fish Responses in Aquaculture. *Entropy* **2014**, *16*, 6133–6151. [CrossRef]

50. Tenorio, B.M.; da Silva, E.A.; Neiva, G.S.M.; da Silva, V.A.; Tenorio, F.D.A.M.; da Silva, T.D.; Silva, E.C.S.E.; Nogueira, R.D. Can fractal methods applied to video tracking detect the effects of deltamethrin pesticide or mercury on the locomotion behavior of shrimps? *Ecotoxicol. Environ. Saf.* **2017**, *142*, 243–249. [CrossRef] [PubMed]

51. Seuront, L.; Cribb, N. Fractal analysis provides new insights into the complexity of marine mammal behavior: A review, two methods, their application to diving and surfacing patterns, and their relevance to marine mammal welfare assessment. *Mar. Mammal Sci.* **2017**, *33*, 847–879. [CrossRef]

52. Seuront, L.; Cribb, N. Fractal analysis reveals pernicious stress levels related to boat presence and type in the Indo-Pacific bottlenose dolphin, *Tursiops aduncus*. *Phys. A-Stat. Mech. Appl.* **2011**, *390*, 2333–2339. [CrossRef]

53. Maximino, C. Modulation of nociceptive-like behavior in zebrafish (*Danio rerio*) by environmental stressors. *Psychol. Neurosci.* **2011**, *4*, 149–155. [CrossRef]

54. Moberg, G.P.; Mench, J.A. *The Biology of Animal Stress: Basic Principles and Implications for Animal Welfare*; CABI Publishing: Wallingford, UK, 2000.

55. Stubsjoen, S.M.; Bohlin, J.; Skjerve, E.; Valle, P.S.; Zanella, A.J. Applying fractal analysis to heart rate time series of sheep experiencing pain. *Physiol. Behav.* **2010**, *101*, 74–80. [CrossRef] [PubMed]

56. Maria, G.A.; Escos, J.; Alados, C.L. Complexity of behavioural sequences and their relation to stress conditions in chickens (*Gallus gallus domesticus*): A non-invasive technique to evaluate animal welfare. *Appl. Anim. Behav. Sci.* **2004**, *86*, 93–104. [CrossRef]

57. Asher, L.; Collins, L.M.; Ortiz-Pelaez, A.; Drewe, J.A.; Nicol, C.J.; Pfeiffer, D.U. Recent advances in the analysis of behavioural organization and interpretation as indicators of animal welfare. *J. R. Soc. Interface* **2009**, *6*, 1103–1119. [CrossRef] [PubMed]

58.    Mason, G.J. Stereotypies and suffering. *Behav. Proc.* **1991**, *25*, 103–115. [CrossRef]
59.    Garner, J.P. Stereotypies and Other Abnormal Repetitive Behaviors: Potential Impact on Validity, Reliability, and Replicability of Scientific Outcomes. *ILAR J.* **2005**, *46*, 106–117. [CrossRef] [PubMed]
60.    Gonyou, H.W. Why the study of animal behavior is associated with the animal-welfare issue. *J. Anim. Sci.* **1994**, *72*, 2171–2177. [CrossRef] [PubMed]
61.    Jasmin, L.; Kohan, L.; Franssen, M.; Janni, G.; Goff, J.R. The cold plate as a test of nociceptive behaviors: Description and application to the study of chronic neuropathic and inflammatory pain models. *Pain* **1998**, *75*, 367–382. [CrossRef]
62.    Deakin, A.G.; Buckley, J.; AlZu'bi, H.S.; Cossins, A.R.; Spencer, J.W.; Al'Nuaimy, W.; Young, I.S.; Sneddon, L.U. Automated monitoring of behaviour in zebrafish after invasive procedures. 2019; MS under review.
63.    Blaser, R.E.; Rosemberg, D.B. Measures of Anxiety in Zebrafish (*Danio rerio*): Dissociation of Black/White Preference and Novel Tank Test. *PLoS ONE* **2012**, *7*, e36931. [CrossRef] [PubMed]
64.    Sneddon, L.U.; Braithwaite, V.A.; Gentle, M.J. Novel object test: Examining nociception and fear in the rainbow trout. *J. Pain* **2003**, *4*, 431–440. [CrossRef]
65.    Sneddon, L.U. Pain in Laboratory Animals: A Possible Confounding Factor? *Altern. Lab. Anim. ATLA* **2017**, *45*, 161–164. [PubMed]
66.    Brown, R.S.; Cooke, S.J.; Anderson, W.G.; McKinley, R.S. Evidence to challenge the '2% rule' for biotelemetry. *N. Am. J. Fish. Manag.* **1999**, *19*, 867–871. [CrossRef]
67.    Ficke, A.D.; Myrick, C.A.; Kondratieff, M.C. The effects of PIT tagging on the swimming performance and survival of three nonsalmonid freshwater fishes. *Ecol. Eng.* **2012**, *48*, 86–91. [CrossRef]
68.    Thorstad, E.B.; Rikardsen, A.H.; Alp, A.; Okland, F. The Use of Electronic Tags in Fish Research—An Overview of Fish Telemetry Methods. *Turk. J. Fish. Aquat. Sci.* **2013**, *13*, 881–896.
69.    Tudorache, C.; Viaene, P.; Blust, R.; Vereecken, H.; De Boeck, G. A comparison of swimming capacity and energy use in seven European freshwater fish species. *Ecol. Freshw. Fish* **2008**, *17*, 284–291. [CrossRef]
70.    Lower, N.; Moore, A.; Scott, A.P.; Ellis, T.; James, J.D.; Russell, I.C. A non-invasive method to assess the impact of electronic tag insertion on stress levels in fishes. *J. Fish Biol.* **2005**, *67*, 1202–1212. [CrossRef]
71.    Fu, C.; Cao, Z.D.; Fu, S.J. The effects of caudal fin loss and regeneration on the swimming performance of three cyprinid fish species with different swimming capacities. *J. Exp. Biol.* **2013**, *216*, 3164–3174. [CrossRef] [PubMed]
72.    Webb, P.W. Effects of median-fin amputation on fast-start performance of rainbow-trout (*Salmo-gairdneri*). *J. Exp. Biol.* **1977**, *68*, 123–135.
73.    Plaut, I. Effects of fin size on swimming performance, swimming behaviour and routine activity of zebrafish *Danio rerio*. *J. Exp. Biol.* **2000**, *203*, 813–820. [PubMed]
74.    Paulus, M.P.; Geyer, M.A.; Sternberg, E. Differential movement patterns but not amount of activity in unconditioned motor behavior of Fischer, Lewis, and Sprague-Dawley rats. *Physiol. Behav.* **1998**, *65*, 601–606. [CrossRef]
75.    Sneddon, L.U. Anatomical and electrophysiological analysis of the trigeminal nerve in a teleost fish, *Oncorhynchus mykiss*. *Neurosci. Lett.* **2002**, *319*, 167–171. [CrossRef]
76.    White, L.J.; Thomson, J.S.; Pounder, K.C.; Coleman, R.C.; Sneddon, L.U. The impact of social context on behaviour and the recovery from welfare challenges in zebrafish, *Danio rerio. Anim. Behav.* **2017**, *132*, 189–199. [CrossRef]
77.    Oliveira, R.F.; Faustino, A.I. Social information use in threat perception: Social buffering, contagion and facilitation of alarm responses. *Commun. Integr. Biol.* **2017**, *10*, e1325049. [CrossRef]
78.    Faustino, A.I.; Tacão-Monteiro, A.; Oliveira, R.F. Mechanisms of social buffering of fear in zebrafish. *Sci. Rep.* **2017**, *7*, 44329. [CrossRef] [PubMed]
79.    Chablais, F.; Jaźwińska, A. Induction of Myocardial Infarction in Adult Zebrafish Using Cryoinjury. *J. Vis. Exp. JOVE* **2012**, *62*, e3666. [CrossRef] [PubMed]
80.    Lemmens, K.; Bollaerts, I.; Bhumika, S.; de Groef, L.; Van Houcke, J.; Darras, V.M.; Van Hove, I.; Moons, L. Matrix metalloproteinases as promising regulators of axonal regrowth in the injured adult zebrafish retinotectal system. *J. Comp. Neurol.* **2016**, *524*, 1472–1493. [CrossRef] [PubMed]
81.    Schweitzer, J.; Becker, T.; Becker, C.G.; Schachner, M. Expression of protein zero is increased in lesioned axon pathways in the central nervous system of adult zebrafish. *Glia* **2003**, *41*, 301–317. [CrossRef] [PubMed]

82. AlZu'bi, H.S. Analysis of Human Activities and Animal Behaviours Based on Computational Intelligence. Ph.D. Thesis, University of Liverpool, Liverpool, UK, 2015.

83. Costa, F.V.; Rosa, L.V.; Quadros, V.A.; Santos, A.R.S.; Kalueff, A.V.; Rosemberg, D.B. Understanding nociception-related phenotypes in adult zebrafish: Behavioral and pharmacological characterization using a new acetic acid model. *Brain Behav. Res.* **2019**, *359*, 570–578. [CrossRef] [PubMed]

84. Sneddon, L.U. Clinical anaesthesia and analgesia in fish. *J. Exot. Pet Med.* **2012**, *21*, 32–43. [CrossRef]

85. Nimkerdphol, K.; Nakagawa, M. 3D locomotion and fractal analysis of Goldfish for acute toxicity bioassay. *Int. J. Biol. Med. Sci.* **2006**, *2*, 180–185.

86. Cooley, J.W.; Tukey, J.W. An Algorithm for the Machine Computation of the Complex Fourier Series. *Math. Comput.* **1965**, *19*, 297–301. [CrossRef]

87. Brand, M.; Granato, M.; Nüsslein-Volhard, C. Keeping and raising zebrafish. *Zebrafish* **2002**, *261*, 7–37.

88. Gunnes, K.; Refstie, T. Cold-branding and fin-clipping for marking of salmonids. *Aquaculture* **1980**, *19*, 295–299. [CrossRef]

89. Readman, G.D.; Owen, S.F.; Murrell, J.C.; Knowles, T.G. Do Fish Perceive Anaesthetics as Aversive? *PLoS ONE* **2013**, *8*, e73773. [CrossRef] [PubMed]

90. EC Severity Assessment. 2018. Available online: http://ec.europa.eu/environment/chemicals/lab_animals/pdf/report_ewg.pdf (accessed on 18 December 2018).

*fishes*

MDPI

*Article*

# Social Behavior and Welfare in Nile Tilapia

Eliane Gonçalves-de-Freitas [1,4,*] , Marcela Cesar Bolognesi [1,4],
Ana Carolina dos Santos Gauy [1,4] , Manuela Lombardi Brandão [1] ,
Percilia Cardoso Giaquinto [2,4] and Marisa Fernandes-Castilho [3,4]

[1]  Departamento de Zoologia e Botânica, Universidade Estadual Paulista (UNESP), Rua Cristóvão Colombo
    2265, São José do Rio Preto, SP 15054-000, Brazil; marcelacbolognesi@gmail.com (M.C.B.);
    ana.gauy@gmail.com (A.C.d.S.G.); manulbrandao@gmail.com (M.L.B.)
[2]  Departamento de Fisiologia, Instituto de Biociências, Universidade Estadual Paulista (UNESP),
    Rua Professor Dr. Antonio Celso Wagner Zanin S/N., Botucatu, SP 18618-689, Brazil;
    percilia.giaquinto@unesp.br
[3]  Departamento de Fisiologia, Setor de Ciências Biológicas, Universidade Federal do Paraná (UFPR),
    Av. Coronel Francisco H. dos Santos 100, Curitiba, PR 81531-970, Brazil; mafernandes@ufpr.br
[4]  Centro de Aquicultura da UNESP, Jaboticabal, SP 14884-900, Brazil
*   Correspondence: eliane.g.freitas@unesp.br; Tel.: +55-17-3221-2375

Received: 7 February 2019; Accepted: 12 March 2019; Published: 27 March 2019

check for updates

**Abstract:** Fish social behavior can be affected by artificial environments, particularly by factors that act upon species that show aggressive behavior to set social rank hierarchy. Although aggressive interactions are part of the natural behavior in fish, if constant and intense, such interactions can cause severe body injuries, increase energy expenditure, and lead the animals to suffer from social stress. The immediate consequence of these factors is a reduced welfare in social fish species. In this paper, we consider the factors that impact on the social behavior and welfare of Nile tilapia, an African cichlid fish widely used both in fish farms and in research; this species is frequently used as a model for physiology and behavior research. This is a polygynous species whose males interact aggressively, establishing a territorial based hierarchy, where a dominant male and several subordinate males arise. When social stability is shrunk, the negative effects of prolonged fighting emerge. In this paper, we summarized how some of the common practices in aquaculture, such as classifying individuals by matching their sizes, water renewal, stock density, and environment lighting affect Nile tilapia social aggressive interactions and, in turn, impact on its welfare. We also discuss some ways to decrease the effects of aggressive interactions in Nile tilapia, such as environment color and body tactile stimulation.

**Keywords:** aggressive interaction; social stress; fighting ability; social rank; social communication

---

## 1. Introduction

Social behavior is defined as any kind of interaction between conspecifics, in such a way that it influences their immediate or future behavior [1]. In this conceptual framework, fish social behavior includes reproductive behavior, such as mating and courtship, cooperative interactions [2,3], shoaling [4], and social hierarchy, which is marked by aggressive interactions to reach a certain social rank and to defend territory, as well as environmental resources [5]. Social interaction, however, is not limited to conspecifics, but it can also occur among heterospecifics, such as in client-cleaner coral reef fishes, wherein a species cooperates by cleaning parasites off of another species, thus interacting between them [6,7].

For each kind of social interaction, several types of information are exchanged between interactants, either directly or indirectly. In the first and more generalized case, information is exchanged between interactants, for instance, during contests to achieve a dominant rank [8]. In the second one, an individual gathers information indirectly from other interactants, and uses that information at a later time, in a similar circumstance. For example, a male *Betta splendens* will quickly engage in a contest after obtaining information about its rival's fighting outcome if it is a loser, but will approach more slowly if facing a winner from a previous fight [9]. Thus, socially organized individuals receive and send several types of messages from/to conspecifics, within a social communication network; it is interpreted depending on several intrinsic and extrinsic factors. When the external environment is changed, social communication can be impaired and social interactions can be affected. For instance, for some fish species chemical information is important in social rank communication [10], however, water dilution can dilute such information, thus, disturbing social stability and increasing aggressive interactions [11,12].

According to Creel et al. [13], social environment is one of the main sources that induces physiological stress in vertebrates, known as social stress. The type of interaction alongside the social rank stimulates the hypothalamic–pituitary–adrenal (HPA) axis, thus increasing the secretion of glucocorticoids, which will impact upon the growth, reproduction, and immune function of animals [13]. Cortisol levels, for instance, are higher for dominant or subordinate individuals, depending on the circumstances [13]. Social instability, in turn, is usually associated to increased stress and its negative effects upon individuals [14,15], as that is a condition characterized by intense and prolonged fighting; therefore, knowing the factors that affect social behavior is an important tool to understand the mechanisms modulating animal welfare.

A rich behavioral repertoire is shown by fish species from the cichlid group, whose social behavior is complex and, although some variations can occur, it does have some patterns as a species. For example, all cichlid species take care of their broods, which can be defined as a bi-parental care, with both the male and the female caring for the eggs and fingerlings in the substrate [16]; or mono-parental care, with only the male or the female taking care of eggs by keeping the fingerlings inside their mouths (mouthbrooding cichlids) [17]. Some species show cooperative behavior as helpers in the group [3]. Despite these behavioral variations, all cichlid species engage in aggressive interactions to establish social rank and territory [16], which is marked by biting, mouth fighting, tail beating (known as overt fight), and by signals such as threats and other displays (restrained aggression). This type of interaction is not only observed in adult males and females, but also in juveniles, which show a very similar aggressive behavior to that of adults. For example, the angelfish *Pterophylum scalare* and the *Cichlasoma paranaense* show a similar social interaction, their social rank being clearly established when they are juveniles [12,18]. Aggressive behavior is, therefore, part of the competition for resources, for example, food, reproductive partners, spawning and brood care site, in which dominant individuals have priority over others to access such resources [19]. Overall, the adaptive value of such social hierarchy is to reduce the detrimental effects of competition, by organizing the access to environmental resources and reducing the cost of prolonged fighting, such as energy expenditure, body injuries, and social stress for the contestants.

Although aggressive interactions are part of their natural behavior, some of the common practices in aquaculture, such as classifying individuals by matching their size, water renewal, stock density, and environment lighting can affect social interactions in a way that shrinks the natural adaptive value of social behavior. As a consequence, there will be reduced welfare in social fish species, as well as an impact on fish production [5]. In this paper, we present a synthesis on social impairment and welfare of Nile tilapia, *Oreochromis niloticus* (Linnaeus, 1758), an African cichlid fish widely used both in fish farms and as a research model due to its physiology and behavior. In fact, tilapia production has grown exponentially over the last few years (global production was approximately 5.6 million tons in 2015, [20]) and it has become the second most farmed fish species worldwide [20]. Thus, it is necessary to consider factors affecting the species' welfare to propose adequate technology for the improvement of tilapia farming and housing.

## 2. Nile Tilapia Social Behavior

Nile tilapia is a polygynous species whose male individuals aggressively interact establishing a territorially based hierarchy [21]. The contestant males fight each other, a winner emerges (dominant male) and starts to defend a territory wherein a circular nest is dug and courtship and spawning will take place [22,23]. Many subordinate males defend territories near the dominant one, although some males do not get any territory at all. Mating occurs through a lek system, with females visiting several nests before mating (Figure 1). After spawning, females take the eggs in their mouths and carry out mouthbrooding for nearly 14 days, until broods are completely released in the open environment [21].

Similarly to other cichlids, Nile tilapia adult males and females show social-rank based interactions [24,25]. Juveniles also have a similar behavior to that of adults (e.g., [26,27]), however, it is probably due to a competition for food rather than for reproduction [28]. Depending on the context, social rank is kept among males through physical limits on the ground (territory and nest), as well as through different types of sensorial cues, such as visual [29,30], chemical [31], and acoustic ones [32,33] which counteract overt fights. When some of these social signs are impaired by environmental changes, specifically artificial ones, social rank signals become unable to keep social stability, thus increasing the negative effects of prolonged fighting, such as increased social stress and decreased growth, which are indicated and discussed hereafter.

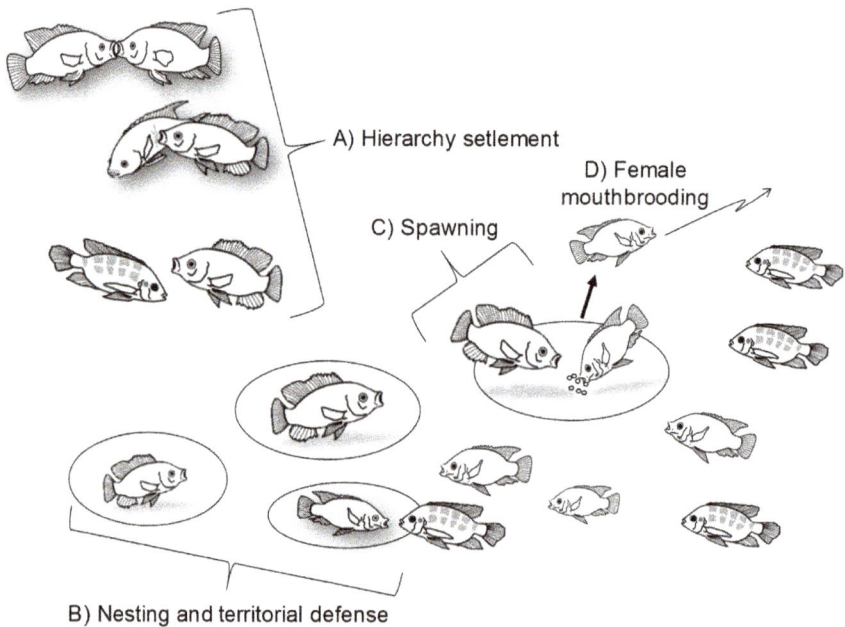

**Figure 1. Nile tilapia's social behavior.** Males fight (A) and a dominant and subordinate rank (clear and gray fish, respectively) emerge; the dominant male starts to defend a territory, wherein it digs a circular nest (B). The mating system is a lek polygyny; then several males dig nests surrounding dominants (B). Fish continue defending territory and attracting females (C). After spawning, the female leaves the arena and starts mouthbrooding until fryers arise (D). Some males do not get territory and stay around the arena challenging territorial males (some act as sneakers). Females (the smallest fish in the drawing) visit the arena and mate with territorial males.

## 3. Social Stress and Socially Controlled Growth

The stress caused by the social organization of animals is an inherent biological characteristic to many animal species, with representative examples on fish and other vertebrate classes [13]. For these species, different social statuses in a group will impinge different demands on individuals and, therefore, different amounts of energy availability resulting in fish with different growth rates and developments, including reproduction. Nile tilapia is one of these species, whose causal factors of intraspecific growth heterogeneity were extensively investigated by our research group, at the time led by Dr. Gilson Volpato, as shown hereafter.

Although results varied genetically for individuals with different growth rates, we observed that the growth heterogeneity in Nile tilapia is associated with the social rank of the animals in the group; i.e., dominant animals grow more than subordinate ones [34]. Volpato and Fernandes [35] proposed a diagrammatic view of the mechanisms involved in the social control of growth in fish, considering species that present or do not present a hierarchical rank. According to them, the heterogeneous growth found in socially organized fish may be due to a combination of the following factors related to social stress: different food intake rates (paradigm of food competition), different digestion rates, appetite suppression, or different rates of energy expenditure of animals occupying different social statuses in the group (paradigm of social stress). Studies conducted in our laboratory suggest that dominant and subordinate Nile tilapia individuals eat equivalent amounts when food is not a limited resource, with a temporal difference in intake; dominant animals eat before subordinate ones [35,36]. The stress arising from the social hierarchy of dominance is considered to be the main promoter of growth heterogeneity in Nile tilapia. Subordinate individuals show higher metabolic rates than dominant ones, thus indicating that most of the energy is used for other purposes. Studies developed by Fernandes and Volpato [37] on the effect of social stress on carbohydrate metabolism in adult Nile tilapia were pioneers in demonstrating that subordinate animals had a glycemic concentration that was twice as high as dominant ones after two days of grouping, and had a significant decrease in hepatic glycogen concentration after four days, which indicates that subordinates use more energy for adjustment to social stress than dominants.

Metabolic differences between fingerlings from different social ranks were also demonstrated by Alvarenga and Volpato [27], when they found a significant association between some agonistic profiles and metabolism in juvenile Nile tilapia, inferred from oxygen consumption, resistance to progressive hypoxia, and ventilatory rate. The authors pointed out that the metabolic variability among individuals of the same social status is directly related to the agonistic profile of male interactants. Then, the individual growth rate in Nile tilapia results from the individual metabolic pattern impinged by the stress caused by the social hierarchical status. Subordinates grow less than dominants, in a linear scale, according to their social rank.

Recently, de Verdal et al. [28] found no correlation between social rank, food conversion efficiency, and growth rates in juvenile Nile tilapia. The authors concluded that such a result could be explained by a low number of agonistic interactions and the fact that fish were under low food competition. However, according to Carrieri and Volpato [36], the snatching frequency is not a reliable parameter to indicate an individual food intake among Nile tilapia, which problematizes studies that require the evaluation of the cumulative effect of competition on food intake, such as growth or conversion efficiency studies. On the other hand, de Verdal et al. [28] studied groups of 15 individuals, which suggests that social hierarchy could be affected by the number of animals interacting in the group.

## 4. Impacts on Social Rank-Based Behavior

### 4.1. Body Size and Fighting Abilities

An important cue related to social information is the access to the opponent's fighting abilities [38]. In this sense, body size constitutes crucial information, as several cichlid species can visually access this characteristic in their opponents (see [39], for *Melanochromis auratus*, and [40] for *Astronotus ocellatus*). The more similar the size, the longer and harder the fighting, as shown for *Nannacara anomala* [41]. Despite this knowledge, fish like Nile tilapia are selected according to their similar size during grading management in aquaculture systems [42], which results in fish with similar fighting abilities being in the same tank. This method can increase chances of mortality as a consequence of massive energy expenditure and also of physical injuries. In fact, this is the case with Nile tilapia. Boscolo et al. [43] studied the effect of matching Nile tilapia males, GIFT lineage, according to their size. They compared aggressive interactions in groups formed by five homogeneously sized or heterogeneously sized males. They found that fish had twice as many fights in the homogeneous group compared to the heterogeneous one; they also showed social instability, although the cortisol level was similar in both treatments. Furthermore, Barreto et al. [44], found a similar result for Nile tilapia, Thai lineage, in which fish showed increased body lesions and scales loss when grouped according to their size. In this case, cortisol was equally elevated for every individual in the similar-sized group, whereas it was high only for alpha and beta fish in the heterogeneous group. Altogether, these studies clearly demonstrate the negative impact of gathering fish with similar sizes (and similar fighting abilities) on the Nile tilapia welfare. As an example, Garcia et al. [45] found a higher growth performance for Nile tilapia in tanks where no size selection was applied. Thus, the heterogeneous growth originated from social rank system seems to be an adaptive mechanism for reducing overt fights in Nile tilapia, indicating that grading should be rethought.

### 4.2. Stocking Density

Stocking density, *strictu sensu*, is the concentration at which fish are initially stocked into a system, but the term has been used to refer to the density of fish at any point in time, considering either their biomass or the number of fish [46]. The stocking density is directly linked to welfare as it affects food competition and consumption, growth, stress, health, and mortality [46]. For social species, the number of individuals in a group is associated to the probability of encounters. As a result, we would expect that the larger the group is, the higher the probability of fighting. However, the contrary is observed for other species, such as salmonids, whose aggressive interactions reduce at high stocking densities [46,47].

There are several studies regarding the effects of stocking density on the Nile tilapia production, but only a few suggesting the effects of social behavior, although they do not quantify aggressive interactions. For instance, according to Ellison et al. [48], Nile tilapia reared at low stocking densities (fry LD = 94 individuals at 1.5 kg m$^{-3}$ vs. fry HD = 366 individuals at 6 kg m$^{-3}$) have an increased expression of genes related to stress which is likely due to increased aggressive interactions; moreover, they are more susceptible to the consequences of infection by *Saprolegnia parasitica*, and have higher mortality rates. Overall, these authors showed that Nile tilapia reared at low densities have a higher susceptibility to negative effects than those reared at higher densities. On the other hand, Garcia et al. [45] showed that a low stocking density (130 juveniles m$^{-3}$ vs. 450 juveniles m$^{-3}$) improves growth rate and food conversion in Nile tilapia. Among several factors, the authors discussed the probability of a better opportunity for both subordinate and dominant fish to access food and to reduce aggressive interactions among them. Barcellos et al. [49] studied the Nile tilapia's response to an acute stressor (net chasing) at four stocking densities (groups of one, two, five, and 10 fish/100 L$^{-1}$). They found higher cortisol levels at stocking densities of 10 fish compared to those of one, two, and five. They attributed these results to detrimental effects from social stress, which was probably caused by agonistic interactions. Although these three studies discuss the effects of social behavior on Nile tilapia's performance under different stock densities, none of them quantified their social interactions. Indeed, the number of animals clearly hinders the aggressive behavior quantification for most of those conditions. Furthermore, there are differences in the experimental protocol between those studies that make it difficult to conclude only in terms of high or low density if social hierarchy can be affected by the number of animals in the group, without considering the amount of available food, shelter, and space [5]. Nevertheless, studies combining the effect of stocking density and social aggressive interactions should be conducted in order to allow us a better understanding of its impact on Nile tilapia's welfare.

### 4.3. Chemical Communication and Social Rank Signaling

Cichlids are known to use chemical information in several social interactions, particularly when signaling social rank, as in *Astatotilapia burtoni* [50], for example. In fact, differences between dominants and subordinates have been shown for the well-studied species Mozambique tilapia *Oreochromis mossambicus* [10,51], whose male stores urine and releases it when a male intruder invades its territory [52]. Furthermore, males reduce their aggressiveness against a mirror when exposed to the urine of a dominant male [53], showing that urinary odorants act as a dominance signal for this species, modulating the aggressiveness in rival males and keeping the social hierarchy stable.

The chemical communication is also important for the social behavior of this species. The first authors to show the role of chemical communication in Nile tilapia were Giaquinto and Volpato [31]. They observed that the recognition of social positions for such species was impaired when chemical communication was absent, even when there was visual contact. Male pairs were kept in an aquarium and were separated by a transparent glass (so that they had visual contact) with a hole in it, allowing water to flow between compartments. When the hole was capped and chemical communication was interrupted, fish spent more time fighting and hierarchy was not established. These results showed that chemical cues can play an important role in Nile tilapia's dominance relationships. Later, Gonçalves-de-Freitas et al. [11] simulated a scenario of water changing, a common practice in fish maintenance. They tested the effect of water renewal on aggressive interactions between pairs of juvenile male Nile tilapia, and found that after 50% water changing, subordinate fish started to fight dominant ones again, destabilizing social hierarchy. This showed that chemicals which signalized dominance had been washed out through water renewal, stimulating fights in the group [11]. How, then, could we control the water quality without washing away important social signals? The answer could be the amount of water renewed. In the juvenile angelfish cichlid, *Pterophyllum scalare*, for example, a change of 25% water stimulated aggressive interactions in the group for one hour, and after this period aggressive interactions returned to basal levels [12]. On the other hand, after changing 50% water, fish kept an increased aggressive interaction rate for 24 h [12]. Thus, it is possible that other cichlid species, such as Nile tilapia, also behave that way, although this hypothesis needs to be tested.

### 4.4. Environment Lighting

Changes in environment lighting can affect aggressive interactions and stress in fish by increasing light intensity [54] or by changing the daily photoperiod [55]. These effects are usually indirect, since they change the levels of melatonin in individuals, which is a light-controlled hormone involved in the modulation of several types of behavior, such as biological rhythm, color change, and aggressiveness [56]. High levels of melatonin are associated to reduced aggressive behaviors in fish and other vertebrates, whereas the opposite is also true, i.e., reduced melatonin is associated with increased aggressive behaviors [57–60]. Thus, long days and high intensity illumination are expected to increase aggressiveness, although some differences can be found among fish species and developmental stages. Environment lighting is a matter of consideration for the Nile tilapia welfare, because they have a good sense of vision [61,62]; consequently, light can affect their behavior and physiology in several ways. These factors are, however, more relevant in aquaria systems, inside laboratories, where the environment is more translucent than in ponds.

In Nile tilapia, increased light intensity (from 280 to 1390 lx) reduces aggressive interactions between pairs of juvenile males [63], whereas it clearly increases fights among adult males [64]. Although it is known that cichlids are more aggressive and tend to become dominant under long photoperiods, such as *Cichlasoma dimerus* [65] and *Tilapia rendalli* [55], information about this influence on Nile tilapia's social behavior is scarce in the literature. According to Martinez-Chavez et al. [66], Nile tilapia shows a daily variation in melatonin levels, which follows a clock-controlled rhythm (low during photophase and high during scotophase); it allows us to hypothesize that the day length can affect social aggressive interactions as it happens for other cichlids. Studies regarding the effect of day length on Nile tilapia is related to other important parameters to aquaculture purposes, such as feeding rate, growth, and reproduction [67]. Thus, the effect of day length on aggressive interactions of Nile tilapia is still an open research field.

## 5. Mitigation Technics

A way to reduce the detrimental effects of fish fights is, obviously, to know what increases aggressive encounters and, then, shape the artificial environment accordingly. For example, finding ways to keep low intensity illumination in translucent environments, renewing a lower amount of water from the tanks, or avoiding grouping fish with similar fighting abilities (same sized fish) in the same place. Knowledge on some of these mitigating factors is still scarce or confused, such as stocking density. Therefore, it is obviously necessary to keep studying in order to find out how external environment and management practices affect the social behavior of Nile tilapia. However, there are some other strategies that can help reduce aggressive interactions. We will now discuss some of them.

### 5.1. Environmental Enrichment

The first solution we can arrive at in order to improve Nile tilapia's welfare is providing them with a more enriched environment. As a definition, environmental enrichment is a modification of the environment, which may increase the animals' behavioral opportunities and lead to improvements in their biological function [68]. Enrichment offers new opportunities to express behaviors and to achieve preferred activities by animals [69,70]. In addition, it may reduce abnormal behaviors, such as stereotypies [69].

Some studies with fish had already shown that environmental enrichment can reduce the behavioral deficit in poor artificial environments [71–73]. It was also demonstrated that enrichment decreases aggression rates between individuals in zebra fish (*Danio rerio*) [74], in the convict cichlid (*Archocentrus nigrofasciatus*, redescribed as *Amatitlania nigrofasciata*) [75], in the pearl cichlid (*Geophagus brasiliensis*) [76], and in the redbreast tilapia (*Tilapia rendalli*) [77]. These authors discuss that an enriched and more complex environment may decrease the probability of encounters between animals, by reducing the visibility between opponents or limiting the boundaries of small territories in the environment, leading to a decrease in aggressive interactions in these species. However, the opposite occurs for Nile tilapia. According to Barreto et al. [78], environmental enrichment increases the aggressiveness in Nile tilapia males. Such an effect occurs because enrichment increases the value of the resource (territory). Thus, individuals are more aggressive when striving for valuable resources to themselves. In fact, the cost of fight can increase as the resource value increases [79]. Therefore, environmental enrichment is not a way to counteract aggressive interactions in Nile tilapia, and would reduce welfare instead.

### 5.2. Body Tactile Stimulation

Tactile stimulation is achieved through a mechanical stimulus, such as touches on the body, when two or more individuals are interacting [80], or artificially, by using some devices [81,82]. Such stimulation has been studied as an alternative to promote animal welfare in several animal groups. In mammals, for example, tactile stimulation reduces both stress levels [83] and heart beat rates [84]. Soares et al [81] demonstrated that tactile stimulation decreases stress in a coral reef fish, decreasing its cortisol levels after confinement. However, Bolognesi et al. [82] demonstrated that tactile stimulation does not decrease the cortisol level in Nile tilapia immediately after the application of a stressor (either social or non-social stressor), but it reduces the aggressiveness in pair staged fights. Such an effect was supposed to be controlled by serotonin (5HT), since this neurotransmitter is released during body tactile stimulation and acts as an inhibitor of aggression in fish [85,86]. This present study raises the potential of using tactile stimulation as a way to mitigate the negative effects of social aggressive interactions on Nile tilapia and other fish species.

## 5.3. Environment Color

Fish have photoreceptor cones in their eyes, which give them color vision [87,88], and allow them to discriminate between different colors in the environment [89]. The most common methods for studying the effects of the colors of the environment are through testing different background colors [90] or applying colors to the environment [91]. A red environment, for example, stimulates food intake but not growth [92], whereas a blue environment is efficient to improve reproduction in Nile tilapia [93]. Additionally, a blue environment influences Nile tilapia's HPI axis, since isolated fish show lower cortisol levels after a confinement stress [94]. Maia and Volpato [91] also demonstrated that a blue lighted environment prevents the increase in the ventilatory frequency in confined fish. Although how the color of the environment will affect Nile tilapia's social aggressive behavior is not available in the literature, we assume that this could be a way to reduce the detrimental effects of aggressive interactions, since the blue color reduces stress levels in Nile tilapia.

## 5.4. Preference Tests

Naturally Nile tilapia's welfare is not limited to their social aggressive behavior. Even though it is not so easy to reduce this inherent individual stressor, it is possible to adopt several measures upon fish behavior. Preference tests, for instance, are widely used to promote better animal welfare; through them, we know what an animal wants regardless of physiological indicators [95–97]. By knowing what animals prefer, we can shape the environment by providing the preferred items and therefore, increase the fish welfare level. However, preference tests should be carried out under several conditions, taking into account successive choices among several items, and keeping in mind that preference may vary from individual to individual and over time [89,95]).

Some preference tests were carried out with Nile tilapia over the last years. For example, Luchiari et al. [98] demonstrated that individuals prefer to access places lit by yellow light rather than red, green, blue, or white light. However, a recent study conducted by Maia and Volpato [99] showed that it takes at least 10 days of testing to find the color preference for this species, green and blue being the most preferred colors in the population. Another study assessing the motivation of Nile tilapia to access places with distinct colors also showed that they were less motivated to access yellow and red, but were more motivated to access green and blue [89]. Few individuals prefer yellow, which reinforces individual preference in such tests [89].

Mendonça et al. [100] showed that Nile tilapia males chose to make their nests in sand substrate when compared to other substrates such as stones; this indicates that the weight and structure of the substrate is evaluated by fish. Furthermore, Freitas and Volpato [101] tested different substrates granulometry and also demonstrated that tilapia individuals prefer small-grained substrates. This preference was more consistent during the morning period than the afternoon, proving the relevance of the period of the day. As we can see, preference tests are still scarce to evaluate Nile tilapia welfare. Further research is needed to understand how preferences could be associated to aggressive interactions and if such preferences are capable of reducing their consequent detrimental effects.

## 6. Concluding Remarks

Several studies regarding the welfare in Nile tilapia consider growth, stress, and reproduction. However, some of the problems found in aquaculture and the welfare of tilapia are, in fact, consequences of their social interactions. It is necessary to keep in mind that social aggressive interactions depend on the external environment, as well as on fish communication, and social environment, which includes information and response from other individuals in a group (Figure 2).

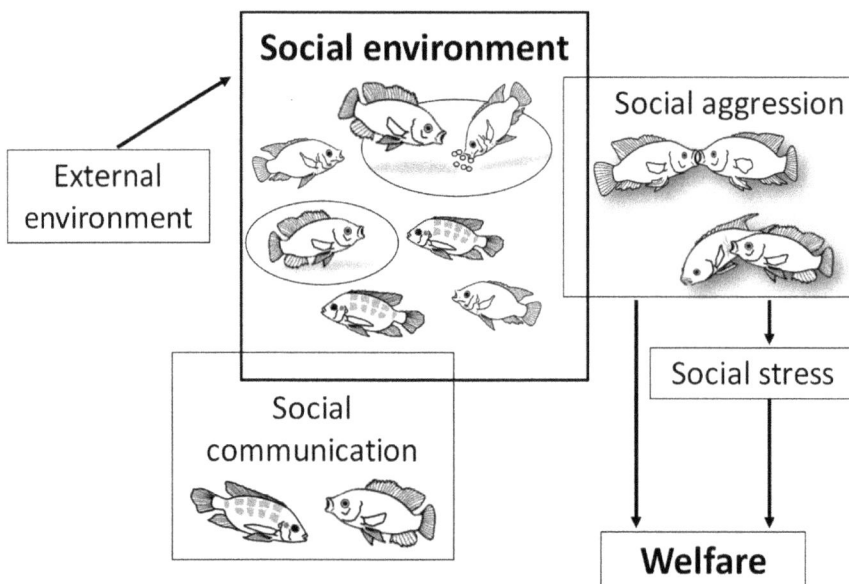

**Figure 2. Summary of factors in artificial environment that would affect Nile tilapia's social behavior.** For instance, light intensity, color, water renewal, grading for size, stocking density, and any other management that directly affects social communication. When such changes increase aggressive interactions, the welfare will be impaired by increased social stress levels and other effects from fighting, such as body injuries and very high energy expenditure. In summary, everything that affects social environment will affect welfare as well.

What is necessary therefore to know about social behavior in order to deal with Nile tilapia welfare? The main points are listed here:

1.  Knowing communication channels and their role as social signaling to keep a good environment for social communication. Lack of information on social rank will increase aggressive interactions.
2.  The dynamics of social interaction, including information regarding the adaptive value of such behavior (e.g., fighting ability). In this sense, grading to gather same sized fish will group together individuals with the same fighting abilities. The consequence will be increased aggressions and, certainly, mortality will be higher than that already considered during fish management.
3.  Factors affecting aggressiveness (such as environment lighting and color) and physiological control, both neural and hormonal, of individual aggressiveness. Adequate control on such variables will help not to exaggerate the effects from aggressiveness.
4.  Strategies to counteract aggressive behaviors, such as body tactile stimulation, which will increase welfare and health in several ways. This is still a developing possibility for fish keeping and aquaculture.

**Author Contributions:** E.G.-d.-F. conceived the idea for the revision and organized the topics to be covered. Since this article is a review, all authors contributed by providing data information, writing and critically reading the content. All authors agree with the ideas and conclusions herein.

**Funding:** EGF research is supported by the "National Council of Technological and Scientific Development"—CNPq (# 310648/2016-5).

**Acknowledgments:** We acknowledge the main person responsible for these 30 years of history regarding Nile tilapia behavior and welfare: Gilson L. Volpato—now at IGVEC (Gilson Volpato's Institute for Scientific Education).

**Conflicts of Interest:** There is no conflict of interest.

# References

1. Robinson, G.E.; Fernald, R.D.; Clayton, D.F. Genes and Social Behavior. *Science* **2008**, *322*, 896–900. [CrossRef]
2. Teresa, F.B.; Gonçalves-de-Freitas, E. Reproductive Behavior and Parental Roles of the Cichlid Fish *Laetacara araguaiae*. *Neotrop. Ichthyol.* **2011**, *9*, 355–362. [CrossRef]
3. Tanaka, H.; Frommen, J.G.; Koblmüller, S.; Sefc, K.M.; McGee, M.; Kohda, M.; Awata, S.; Hori, M.; Taborsky, M. Evolutionary Transitions to Cooperative Societies in Fishes Revisited. *Ethology* **2018**, *124*, 777–789. [CrossRef]
4. Kasumyan, A.O.; Pavlov, D.S. Evolution of Schooling Behavior in Fish. *J. Ichthyol.* **2018**, *58*, 670–678. [CrossRef]
5. Damsgård, B.; Huntingford, F. Fighting and Aggression. In *Aquaculture and Behavior*; Huntingford, F., Jobling, M., Kadri, S., Eds.; Wiley-Blackwell: Oxford, UK, 2012; p. 340.
6. Paula, J.R.; Messias, J.P.; Grutter, A.S.; Bshary, R.; Soares, M.C. The Role of Serotonin in the Modulation of Cooperative Behavior. *Behav. Ecol.* **2015**, *26*, 1005–1012. [CrossRef]
7. Triki, Z.; Bshary, R.; Grutter, A.S.; Ros, A.F.H. The Arginine-Vasotocin and Serotonergic Systems Affect Interspecific Social Behaviour of Client Fish in Marine Cleaning Mutualism. *Physiol. Behav.* **2017**, *174*, 136–143. [CrossRef]
8. Desjardins, J.K.; Fernald, R.D. How Do Social Dominance and Social Information Influence Reproduction and the Brain? *Integr. Comp. Biol.* **2008**, *48*, 596–603. [CrossRef] [PubMed]
9. Oliveira, R.F.; McGregor, P.K.; Latruffe, C. Know Thine Enemy: Fighting Fish Gather Information from Observing Conspecific Interactions. *Proc. R. Soc. B Biol. Sci.* **1998**, *265*, 1045–1049. [CrossRef]
10. Keller-Costa, T.; Canário, A.V.M.; Hubbard, P.C. Chemical Communication in Cichlids: A Mini-Review. *Gen. Comp. Endocrinol.* **2015**, *221*, 64–74. [CrossRef] [PubMed]
11. Gonçalves-de-Freitas, E.; Teresa, F.B.; Gomes, F.S.; Giaquinto, P.C. Effect of Water Renewal on Dominance Hierarchy of Juvenile Nile Tilapia. *Appl. Anim. Behav. Sci.* **2008**, *112*, 187–195. [CrossRef]
12. dos Santos Gauy, A.C.; Boscolo, C.N.P.; Gonçalves-de-Freitas, E. Less Water Renewal Reduces Effects on Social Aggression of the Cichlid *Pterophyllum scalare*. *Appl. Anim. Behav. Sci.* **2018**, *198*, 121–126. [CrossRef]
13. Creel, S.; Dantzer, B.; Goymann, W.; Rubenstein, D.R. The Ecology of Stress: Effects of the Social Environment. *Funct. Ecol.* **2013**, *27*, 66–80. [CrossRef]
14. Sapolsky, R.M. Cortisol Concentrations and the Social Significance of Rank Instability among Wild Baboons. *Psychoneuroendocrinology* **1992**, *17*, 701–709. [CrossRef]
15. Carvalho, R.R.; Palme, R.; da Silva Vasconcellos, A. An Integrated Analysis of Social Stress in Laying Hens: The Interaction between Physiology, Behaviour, and Hierarchy. *Behav. Processes* **2018**, *149*, 43–51. [CrossRef] [PubMed]
16. Barlow, G. *The Cichlid Fishes: Nature's Grand Experiment in Evolution*, 1st ed.; Basic Books: New York, NY, USA, 2002.
17. Balshine, S.; Sloman, K. Parental Care in Fishes. In *Encyclopedia of Fish Physiology: From Genome to Environment*; Farrel, A.P., Ed.; Academic Press: San Diego, CA, USA, 2011; Volume 1, pp. 670–677.
18. Brandão, M.L.; Colognesi, G.; Bolognesi, M.C.; Costa-Ferreira, R.S.; Carvalho, T.B.; Gonçalves-de-Freitas, E. Water Temperature Affects Aggressive Interactions in a Neotropical Cichlid Fish. *Neotrop. Ichthyol.* **2018**, *16*, 1–8. [CrossRef]
19. Huntingford, F.A.; Turner, A.K. (Eds.) *Animal Conflict*; Springer: Dordrecht, The Netherlands, 1987. [CrossRef]
20. FAO. *World Aquaculture 2015: A Brief Overview*; FAO: Rome, Italy, 2017; Volume 1140.
21. Lowe-McConnell, R.H. Observations on the Biology of *Tilapia nilotica* Linné in East African Waters. *Rev. Zool. Bot. Afr.* **1958**, *57*, 129–170.
22. Gonçalves-de-Freitas, E.; Nishida, S.M. Sneaking Behavior of the Nile Tilapia. *Bol. Técnico CEPTA* **1998**, *11*, 71–79.
23. Mendonça, F.Z.; Gonçalves-de-Freitas, E. Nest Deprivation and Mating Success in Nile Tilapia (Teleostei: Cichlidae). *Rev. Bras. Zool.* **2008**, *25*, 413–418. [CrossRef]
24. Carvalho, T.; Gonçalves-de-Freitas, E. Sex Group Composition, Social Interaction, and Metabolism in the Fish Nile Tilapia. *Braz. J. Biol.* **2008**, *68*, 807–812. [CrossRef]

25. Gonçalves-de-freitas, E.; Ferreira, A.C.; Paulista, U.E.; José, S. Female Social Dominance Does Not Establish Mating Priority in Nile Tilapia. *Etologia* **2004**, *6*, 33–37.

26. Pinho-Neto, C.F.; Miyai, C.A.; Sanches, F.H.C.; Giaquinto, P.C.; Delicio, H.C.; Barcellos, L.J.G.; Volpato, G.L.; Barreto, R.E. Does Sex Influence Intraspecific Aggression and Dominance in Nile Tilapia Juveniles? *Behav. Process.* **2014**, *105*, 15–18. [CrossRef] [PubMed]

27. Alvarenga, C.M.D.; Volpato, G.L. Agonistic Profile and Metabolism in Alevins of the Nile Tilapia. *Physiol. Behav.* **1995**, *57*, 75–80. [CrossRef]

28. de Verdal, H.; O'Connell, C.M.; Mekkawy, W.; Vandeputte, M.; Chatain, B.; Bégout, M.-L.; Benzie, J.A.H. Agonistic Behaviour and Feed Efficiency in Juvenile Nile Tilapia *Oreochromis niloticus*. *Aquaculture* **2019**, *505*, 271–279. [CrossRef]

29. Falter, U. Description of the Color Patterns in *Oreochromis niloticus* (L) (Teleostei, Cichlidae). *Ann. Soc. R. Zool. Belgique* **1987**, *117*, 201–219.

30. Volpato, G.L.; Luchiari, A.C.; Duarte, C.R.A.; Barreto, R.E.; Ramanzini, G.C. Eye Color as an Indicator of Social Rank in the Fish Nile Tilapia. *Braz. J. Med. Biol. Res.* **2003**, *36*, 1659–1663. [CrossRef] [PubMed]

31. Giaquinto, P.C.; Volpato, G.L. Chemical Communication, Aggression, and Conspecific Recognition in the Fish Nile Tilapia. *Physiol. Behav.* **1997**, *62*, 1333–1338. [CrossRef]

32. Longrie, N.; Poncin, P.; Denoël, M.; Gennotte, V.; Delcourt, J.; Parmentier, E. Behaviours Associated with Acoustic Communication in Nile Tilapia (*Oreochromis niloticus*). *PLoS ONE* **2013**, *8*, e61467. [CrossRef] [PubMed]

33. Longrie, N.; Van Wassenbergh, S.; Vandewalle, P.; Mauguit, Q.; Parmentier, E. Potential Mechanism of Sound Production in *Oreochromis niloticus* (Cichlidae). *J. Exp. Biol.* **2009**, *212*, 3395–3402. [CrossRef]

34. Volpato, G.L.; Frioli, P.M.A.; Carrieri, M.P. Heterogeneous Growth in Fishes: Some New Data of Nile Tilapia *Oreochromis niloticus* and a General View about the Causal Mechanisms. *Bol. Fisiol. Anim.* **1989**, *13*, 7–22.

35. Volpato, G.L.; Fernandes, M.O. Social Control of Growth in Fish. *Braz. J. Med. Biol. Res.* **1994**, *27*, 797–810.

36. Carrieri, M.P.; Volpato, G.L. Does Snatching Frequency Really Indicate Food Ingestion in the Nile Tilapia? *Physiol. Behav.* **1991**, *50*, 489–492. [CrossRef]

37. Fernandes, M.O.; Volpato, G.L. Heterogeneous Growth in the Nile Tilapia: Social Stress and Carbohydrate Metabolism. *Physiol. Behav.* **1993**, *54*, 319–323. [CrossRef]

38. Enquist, M.; Leimar, O. Evolution of Fighting Behaviour: Decision Rules and Assessment of Relative Strength. *J. Theor. Biol.* **1983**, *102*, 387–410. [CrossRef]

39. Nelissen, M.H.J. Does Body Size Affect the Ranking of a Cichild Fish in a Dominance Hierarchy? *J. Ethol.* **1992**, *10*, 153–156. [CrossRef]

40. Beeching, S.C. Visual Assessment of Relative Body Size in a Cichlid Fish, the Oscar, *Astronotus ocellatus*. *Ethology* **1992**, *90*, 177–186. [CrossRef]

41. Enquist, M.; Ljungberg, T.; Zandor, A. Visual Assessment of Fighting Ability in the Cichlid Fish *Nannacara anomala*. *Anim. Behav.* **1987**, *35*, 1262–1264. [CrossRef]

42. Slavík, O.; Pešta, M.; Horký, P. Effect of Grading on Energy Consumption in European Catfish Silurus Glanis. *Aquaculture* **2011**, *313*, 73–78. [CrossRef]

43. Boscolo, C.N.P.; Morais, R.N.; Gonçalves-de-Freitas, E. Same-Sized Fish Groups Increase Aggressive Interaction of Sex-Reversed Males Nile Tilapia GIFT Strain. *Appl. Anim. Behav. Sci.* **2011**, *135*, 154–159. [CrossRef]

44. Barreto, T.N.; Boscolo, C.N.P.; Gonçalves-de-Freitas, E. Homogeneously Sized Groups Increase Aggressive Interaction and Affect Social Stress in Thai Strain Nile Tilapia (*Oreochromis niloticus*). *Mar. Freshw. Behav. Physiol.* **2015**, *48*, 309–318. [CrossRef]

45. Garcia, F.; Romera, D.M.; Gozi, K.S.; Onaka, E.M.; Fonseca, F.S.; Schalch, S.H.C.; Candeira, P.G.; Guerra, L.O.M.; Carmo, F.J.; Carneiro, D.J.; et al. Stocking Density of Nile Tilapia in Cages Placed in a Hydroelectric Reservoir. *Aquaculture* **2013**, *410–411*, 51–56. [CrossRef]

46. Ellis, T.; North, B.; Scott, A.P.; Bromage, N.R.; Porter, M.; Gadd, D. The Relationships between Stocking Density and Welfare in Farmed Rainbow Trout. *J. Fish Biol.* **2002**, *61*, 493–531. [CrossRef]

47. Adams, C.E.; Turnbull, J.F.; Bell, A.; Bron, J.E.; Huntingford, F.A. Multiple Determinants of Welfare in Farmed Fish: Stocking Density, Disturbance, and Aggression in Atlantic Salmon (*Salmo salar*). *Can. J. Fish. Aquat. Sci.* **2007**, *64*, 336–344. [CrossRef]

48. Ellison, A.R.; Uren Webster, T.M.; Rey, O.; Garcia de Leaniz, C.; Consuegra, S.; Orozco-terWengel, P.; Cable, J. Transcriptomic Response to Parasite Infection in Nile Tilapia (*Oreochromis niloticus*) Depends on Rearing Density. *BMC Genom.* **2018**, *19*, 723. [CrossRef]

49. Barcellos, L.J.G.; Nicolaiewsky, S.; De Souza, S.M.G.; Lulhier, F. Plasmatic Levels of Cortisol in the Response to Acute Stress in Nile Tilapia, *Oreochromis niloticus* (L.), Previously Exposed to Chronic Stress. *Aquac. Res.* **1999**, *30*, 437–444. [CrossRef]

50. Maruska, K.P.; Fernald, R.D. Contextual Chemosensory Urine Signaling in an African Cichlid Fish. *J. Exp. Biol.* **2012**. [CrossRef] [PubMed]

51. Almeida, O.G. Urine as a Social Signal in the Mozambique Tilapia (*Oreochromis mossambicus*). *Chem. Senses* **2005**, *30* (Suppl. 1), i309–i310. [CrossRef] [PubMed]

52. Barata, E.N.; Hubbard, P.C.; Almeida, O.G.; Miranda, A.; Canário, A.V. Male Urine Signals Social Rank in the Mozambique Tilapia (*Oreochromis mossambicus*). *BMC Biol.* **2007**, *5*, 54. [CrossRef] [PubMed]

53. Keller-Costa, T.; Saraiva, J.L.; Hubbard, P.C.; Barata, E.N.; Canário, A.V.M. A Multi-Component Pheromone in the Urine of Dominant Male Tilapia (*Oreochromis mossambicus*) Reduces Aggression in Rivals. *J. Chem. Ecol.* **2016**, *42*, 173–182. [CrossRef]

54. Lopes, A.C.C.; Villacorta-Correa, M.A.; Carvalho, T.B. Lower Light Intensity Reduces Larval Aggression in Matrinxã, *Brycon amazonicus*. *Behav. Process.* **2018**, *151*, 62–66. [CrossRef] [PubMed]

55. Gonçalves-de-Freitas, E.; Carvalho, T.B.; Oliveira, R.F. Photoperiod Modulation of Aggressive Behavior Is Independent of Androgens in a Tropical Cichlid Fish. *Gen. Comp. Endocrinol.* **2014**, *207*, 41–49. [CrossRef]

56. Walton, J.C.; Weil, Z.M.; Nelson, R.J. Influence of Photoperiod on Hormones, Behavior, and Immune Function. *Front. Neuroendocrinol.* **2011**, *32*, 303–319. [CrossRef] [PubMed]

57. Munro, A.D. Effects of Melatonin, Serotonin, and Naloxone on Aggression in Isolated Cichlid Fish (*Aequidens pulcher*). *J. Pineal Res.* **1986**, *3*, 257–262. [CrossRef] [PubMed]

58. Jasnow, A.M.; Huhman, K.L.; Bartness, T.J.; Demas, G.E. Short Days and Exogenous Melatonin Increase Aggression of Male Syrian Hamsters (*Mesocricetus auratus*). *Horm. Behav.* **2002**, *42*, 13–20. [CrossRef]

59. Larson, E.T.; Winberg, S.; Mayer, I.; Lepage, O.; Summers, C.H.; Øverli, Ø. Social Stress Affects Circulating Melatonin Levels in Rainbow Trout. *Gen. Comp. Endocrinol.* **2004**, *136*, 322–327. [CrossRef] [PubMed]

60. Munley, K.M.; Rendon, N.M.; Demas, G.E. Neural Androgen Synthesis and Aggression: Insights From a Seasonally Breeding Rodent. *Front. Endocrinol. (Lausanne)* **2018**, *9*. [CrossRef]

61. Castro, A.L.S.; Gonçalves-de-freitas, E.; Volpato, G.L.; Oliveira, C. Visual Communication Stimulates Reproduction in Nile Tilapia, *Oreochromis niloticus* (L.). *Braz. J. Med. Biol. Res.* **2009**, *42*, 368–374. [CrossRef] [PubMed]

62. do Nascimento Falsarella, L.; Brandão, M.L.; Gonçalves-de-Freitas, E. Fish Adjust Aggressive Behavior to Audience Size with Limited Information on Bystanders' Fighting Ability. *Behav. Process.* **2017**, *142*, 116–118. [CrossRef]

63. Carvalho, T.B.; Ha, J.C.; Gonçalves-de-Freitas, E. Light Intensity Can Trigger Different Agonistic Responses in Juveniles of Three Cichlid Species. *Mar. Freshw. Behav. Physiol.* **2012**, *45*, 91–100. [CrossRef]

64. Carvalho, T.B.; Mendonça, F.Z.; Costa-Ferreira, R.S.; Gonçalves-de-Freitas, E. The Effect of Increased Light Intensity on the Aggressive Behavior of the Nile Tilapia, *Oreochromis niloticus* (Teleostei: Cichlidae). *Zoologia* **2013**, *30*, 125–129. [CrossRef]

65. Fiszbein, A.; Cánepa, M.; Vázquez, G.R.; Maggese, C.; Pandolfi, M. Photoperiodic Modulation of Reproductive Physiology and Behaviour in the Cichlid Fish *Cichlasoma dimerus*. *Physiol. Behav.* **2010**, *99*, 425–432. [CrossRef] [PubMed]

66. Martinez-Chavez, C.C.; Al-Khamees, S.; Campos-Mendoza, A.; Penman, D.J.; Migaud, H. Clock-Controlled Endogenous Melatonin Rhythms in Nile Tilapia (*Oreochromis niloticus niloticus*) and African Catfish (*Clarias gariepinus*). *Chronobiol. Int.* **2008**, *25*, 31–49. [CrossRef]

67. El-Sayed, A.-F.M.; Kawanna, M. Effects of Photoperiod on Growth and Spawning Efficiency of Nile Tilapia (*Oreochromis niloticus* L.) Broodstock in a Recycling System. *Aquac. Res.* **2007**, *38*, 1242–1247. [CrossRef]

68. Newberry, R.C. Environmental Enrichment: Increasing the Biological Relevance of Captive Environments. *Appl. Anim. Behav. Sci.* **1995**, *44*, 229–243. [CrossRef]

69. Mason, G.; Clubb, R.; Latham, N.; Vickery, S. Why and How Should We Use Environmental Enrichment to Tackle Stereotypic Behaviour? *Appl. Anim. Behav. Sci.* **2007**, *102*, 163–188. [CrossRef]

70. van de Weerd, H.A.; Day, J.E.L. A Review of Environmental Enrichment for Pigs Housed in Intensive Housing Systems. *Appl. Anim. Behav. Sci.* **2009**, *116*, 1–20. [CrossRef]
71. Salvanes, A.G.V.; Braithwaite, V.A. Exposure to Variable Spatial Information in the Early Rearing Environment Generates Asymmetries in Social Interactions in Cod (*Gadus morhua*). *Behav. Ecol. Sociobiol.* **2005**, *59*, 250–257. [CrossRef]
72. Salvanes, A.G.V.; Braithwaite, V.A. The Need to Understand the Behaviour of Fish Reared for Mariculture or Restocking. *ICES J. Mar. Sci.* **2006**, *63*, 346–354. [CrossRef]
73. Salvanes, A.G.V.; Moberg, O.; Ebbesson, L.O.E.; Nilsen, T.O.; Jensen, K.H.; Braithwaite, V. a. Environmental Enrichment Promotes Neural Plasticity and Cognitive Ability in Fish. *Proc. Biol. Sci.* **2013**, *280*, 20131331. [CrossRef]
74. Basquill, S.P.; Grant, J.W.A. *An Increase in Habitat Complexity Reduces Aggression and Monopolization of Food by Zebra Fish (Danio Rerio)*; Prentice Hall Inc.: Upper Saddle River, NJ, USA, 1997; Volume 98.
75. Barley, A.J.; Coleman, R.M. Habitat Structure Directly Affects Aggression in Convict Cichlids *Archocentrus nigrofasciatus*. *Curr. Zool.* **2010**, *56*, 52–56.
76. Kadry, V.O.; Barreto, R.E. Environmental Enrichment Reduces Aggression of Pearl Cichlid, *Geophagus brasiliensis*, during Resident-Intruder Interactions. *Neotrop. Ichthyol.* **2010**, *8*, 329–332. [CrossRef]
77. Torrezani, C.S.; Pinho-Neto, C.F.; Miyai, C.A.; Sanches, F.H.C.; Barreto, R.E. Structural Enrichment Reduces Aggression in *Tilapia rendalli*. *Mar. Freshw. Behav. Physiol.* **2013**, *46*, 183–190. [CrossRef]
78. Barreto, R.E.; Carvalho, G.G.A.; Volpato, G.L. The Aggressive Behavior of Nile Tilapia Introduced into Novel Environments with Variation in Enrichment. *Zoology* **2011**, *114*, 53–57. [CrossRef] [PubMed]
79. Arnott, G.; Elwood, R.W. Information Gathering and Decision Making about Resource Value in Animal Contests. *Anim. Behav.* **2008**, *76*, 529–542. [CrossRef]
80. Dunbar, R.I.M. The Social Role of Touch in Humans and Primates: Behavioural Function and Neurobiological Mechanisms. *Neurosci. Biobehav. Rev.* **2010**, *34*, 260–268. [CrossRef]
81. Soares, M.C.; Oliveira, R.F.; Ros, A.F.H.; Grutter, A.S.; Bshary, R. Tactile Stimulation Lowers Stress in Fish. *Nat. Commun.* **2011**, *2*, 534. [CrossRef] [PubMed]
82. Bolognesi, M.C.; dos Santos Gauy, A.C.; Gonçalves-de-Freitas, E. Tactile Stimulation Reduces Aggressiveness but Does Not Lower Stress in a Territorial Fish. *Sci. Rep.* **2019**, *9*, 40. [CrossRef] [PubMed]
83. Tallet, C.; Sy, K.; Prunier, A.; Nowak, R.; Boissy, A.; Boivin, X. Behavioural and Physiological Reactions of Piglets to Gentle Tactile Interactions Vary According to Their Previous Experience with Humans. *Livest. Sci.* **2014**, *167*, 331–341. [CrossRef]
84. Schmied, C.; Waiblinger, S.; Scharl, T.; Leisch, F.; Boivin, X. Stroking of Different Body Regions by a Human: Effects on Behaviour and Heart Rate of Dairy Cows. *Appl. Anim. Behav. Sci.* **2008**, *109*, 25–38. [CrossRef]
85. Clotfelter, E.D.; O'Hare, E.P.; McNitt, M.M.; Carpenter, R.E.; Summers, C.H. Serotonin Decreases Aggression via 5-HT1A Receptors in the Fighting Fish *Betta splendens*. *Pharmacol. Biochem. Behav.* **2007**, *87*, 222–231. [CrossRef]
86. Zubizarreta, L.; Perrone, R.; Stoddard, P.K.; Costa, G.; Silva, A.C. Differential Serotonergic Modulation of Two Types of Aggression in Weakly Electric Fish. *Front. Behav. Neurosci.* **2012**, *6*, 1–10. [CrossRef] [PubMed]
87. Sabbah, S.; Laria, R.; Gray, S.M.; Hawryshyn, C.W. Functional Diversity in the Color Vision of Cichlid Fishes. *BMC Biol.* **2010**, *8*, 133. [CrossRef]
88. Hofmann, C.M.; O'Quin, K.E.; Justin Marshall, N.; Cronin, T.W.; Seehausen, O.; Carleton, K.L. The Eyes Have It: Regulatory and Structural Changes Both Underlie Cichlid Visual Pigment Diversity. *PLoS Biol.* **2009**, *7*. [CrossRef] [PubMed]
89. Maia, C.M.; Volpato, G.L. Preference Index Supported by Motivation Tests in Nile Tilapia. *PLoS ONE* **2017**, *12*, e0175821. [CrossRef]
90. Torres, I.F.A.; da Silva Ferreira, A.; de Souza e Silva, W.; Mesquita, F.O.; Luz, R.K. Effect of Environmental Color on Learning of Nile Tilapia. *Appl. Anim. Behav. Sci.* **2018**, *209*, 104–108. [CrossRef]
91. Maia, C.M.; Volpato, G.L. Environmental Light Color Affects the Stress Response of Nile Tilapia. *Zoology* **2013**, *116*, 64–66. [CrossRef] [PubMed]
92. Volpato, G.L.; Bovi, T.S.; de Freitas, R.H.A.; da Silva, D.F.; Delicio, H.C.; Giaquinto, P.C.; Barreto, R.E. Red Light Stimulates Feeding Motivation in Fish but Does Not Improve Growth. *PLoS ONE* **2013**, *8*, e59134. [CrossRef] [PubMed]

93. Volpato, G.L.; Duarte, C.R.A.; Luchiari, A.C. Environmental Color Affects Nile Tilapia Reproduction. *Braz. J. Med. Biol. Res.* **2004**, *37*, 479–483. [CrossRef] [PubMed]

94. Volpato, G.L.; Barreto, R.E. Environmental Blue Light Prevents Stress in the Fish Nile Tilapia. *Braz. J. Med. Biol. Res.* **2001**, *34*, 1041–1045. [CrossRef] [PubMed]

95. Volpato, G.; Gonçalves-de-Freitas, E.; Fernandes-de-Castilho, M. Insights into the Concept of Fish Welfare. *Dis. Aquat. Organ.* **2007**, *75*, 165–171. [CrossRef] [PubMed]

96. Dawkins, M.S. The Science of Animal Suffering. *Ethology* **2008**, *114*, 937–945. [CrossRef]

97. Volpato, G.L. Challenges in Assessing Fish Welfare. *ILAR J.* **2009**, *50*, 329–337. [CrossRef] [PubMed]

98. Luchiari, A.C.; do Amaral Duarte, C.R.; de Morais Freire, F.A.; Nissinen, K. Hierarchical Status and Colour Preference in Nile Tilapia (*Oreochromis niloticus*). *J. Ethol.* **2007**, *25*, 169–175. [CrossRef]

99. Maia, C.M.; Volpato, G.L. A History-Based Method to Estimate Animal Preference. *Sci. Rep.* **2016**, *6*, 28328. [CrossRef] [PubMed]

100. Mendonça, F.Z.; Volpato, G.L.; Costa-Ferreira, R.S.; Gonçalves-de-Freitas, E. Substratum Choice for Nesting in Male Nile Tilapia *Oreochromis niloticus*. *J. Fish Biol.* **2010**, *77*, 1439–1445. [CrossRef] [PubMed]

101. Freitas, R.H.A.; Volpato, G.L. Motivation and Time of Day Affect Decision-Making for Substratum Granulometry in the Nile Tilapia *Oreochromis niloticus*. *J. Appl. Ichthyol.* **2013**, *29*, 239–241. [CrossRef]

*Review*

# Positive Welfare for Fishes: Rationale and Areas for Future Study

Isabel Fife-Cook and Becca Franks *

Department of Environmental Studies, New York University; New York City, NY 10003, USA; ifc219@nyu.edu
* Correspondence: krf205@nyu.edu

Received: 23 February 2019; Accepted: 7 May 2019; Published: 23 May 2019

check for
updates

**Abstract:** Traditional animal welfare paradigms have focused on maintaining physical health and mitigating negative impacts to wellbeing. Recently, however, the field has increasingly recognized the importance of positive welfare (i.e., mental and physical states that exceed what is necessary for immediate survival) and accordingly introduced manipulations and indicators of positive welfare for use in agriculture, laboratories, and zoos/aquaria. The creation and monitoring of positive welfare requires an in-depth knowledge of species-specific behavior and biology, which necessitates species-specific or, at a minimum, taxa-specific standards. Research on positive welfare in fish is lagging in this regard and therefore merits further consideration. In this paper, we aim to merge what is already known about positive welfare with the existing fish behavior literature to develop a plan of action for fish welfare research that will ultimately contribute to the development of positive welfare standards and assessment strategies for fish. We begin by exploring the origins of positive welfare research and then outline the physical, psychological and species-specific areas of inquiry that can be investigated in fish. In addition to presenting current findings on fish motivation, emotion, potential sources of positive welfare such as fulfillment of motivational urges (establishing agency, engaging in exploration and learning), and play behavior, we also identify promising areas for future research aimed at developing accurate and appropriate indicators of positive welfare in fish.

**Keywords:** animal behavior; fish welfare; positive welfare; welfare enhancement; negative and positive affect; motivation

---

## 1. Introduction

The past two decades have witnessed mounting interest in fish welfare and mounting interest in positive welfare, i.e., mental and physical states that exceed what is strictly necessary for short-term survival. To date, these trends have mostly progressed in parallel, with very little work at the intersection: positive welfare for fishes. Recent empirical (e.g., [1]) and theoretical breakthroughs [2–4], however, suggest that the time is ripe for such investigations; positive welfare can no longer be considered a luxury—it is critical to understanding and creating even minimally-acceptable welfare—and advances in fish behavior and cognition research [5,6] point to many exciting areas for future work.

Historically, animal welfare science has focused on alleviating suffering (chronic stress, pain, etc.) and identifying physiological and behavioral signs of poor welfare, such as frequent illness, elevated blood cortisol and stereotypic behaviors [7]. Fish research has been no exception to this pattern. However, there is now a growing recognition among animal welfare scientists that pursuing the goal of relieving suffering does not lead to a lack of suffering [4,8], thus indicating that even neutral animal welfare necessarily involves positive welfare.

Interestingly, in the early 2000s, human psychology came to a similar conclusion: that psychological science is not complete without understanding positive experiences and positive emotions [9]. This realization resulted in a new field of study (positive psychology) and the central insight that a life without positive experiences is not merely a neutral life, it is a bad life [9]. Positive experiences protect against boredom, languishing, and emotional distress, and positive emotions strengthen long-term well-being, resiliency, and even improve biological outcomes, such as increased cardiovascular health [10]. Similar patterns are now being identified across the animal kingdom [4,11,12] and there is every indication that the same mechanisms function in fish as well.

The aim of this review is thus 2-fold. First, we wish to document the origins of positive welfare research, investigate its assessment parameters and highlight the significance of positive affect in animal husbandry. Second, we seek to explore the potential ways in which these concepts have been and could be applied to fish welfare research. In addressing these issues, we survey the current literature on positive welfare and identify three important areas of consideration for future work on positive fish welfare: (1) environmental inputs that are likely to produce positive welfare for fish, (2) assessing positive welfare in fish utilizing recent developments in emotion research, and (3) play behavior, which, under certain conditions, is consistent with aspects of positive welfare [13] and likely occurs in some species of fish [14].

## 2. A Brief History of Positive Welfare Scholarship

### 2.1. Original Conceptualizations of Positive Animal Welfare

Fraser and Duncan [7] were among the first welfare researchers to propose incorporating measures of positive affect into welfare assessment. They reasoned that if animals experience negative affective states (pain, hunger/thirst, fear, exhaustion, etc.), evolutionary theory suggests they should experience positive affective states as well. According to Fraser and Duncan [7], motivational affective states (MASs) evolved as separate processes to solve two different types of motivational problems; survival situations (what the animal needs) and opportunity situations (what the animal wants/likes). In both scenarios, the animal's affective state and resulting behavioral responses usually result in an evolutionarily advantageous outcome where the net outcome of the action on the animal's fitness is positive. Affective states are triggered by internal or external stimuli and subsequently motivate certain types of behavioral responses which can be positive (pleasure, joy, excitement, fulfillment, etc.) or negative (pain, hunger/thirst, fear, exhaustion, etc.). From an evolutionary standpoint, it is therefore in the animal's best interest to avoid prolonged negative affective states and promote positive states in order to optimize fitness.

Fraser and Duncan [7] suggested that analyzing behavior using their conceptual framework of positive and negative affect is necessary in order to generate a comprehensive assessment of an animal's welfare. Historically, welfare research has largely focused on mitigating prolonged need situations, i.e., those in which the animal is required to perform an action necessary to cope with threats to their survival and the resulting negative affective states, such as fleeing from a predator or searching for water during a drought. While providing relief from need situations will have a net beneficial effect on poor animal welfare, the most this approach can accomplish is to raise the welfare status from poor to neutral. While prolonged need situations are detrimental to animal welfare, their absence does not indicate positive affect. Fraser and Duncan [7] argued that providing animals with situations to promote positive affective states such as pleasure and joy are necessary to improve welfare past the neutral point. This work provided the imperative to incorporate positive welfare markers and use a combination of behavioral analysis and life history knowledge to predict when an aspect of animal management is likely to cause a positive reaction.

While the evolutionary theory and behavioral observations discussed by Fraser and Duncan [7] strongly suggested that positive emotions are not exclusive to humans, there was still little practical information regarding ways to promote positive affect in animals. Yeates and Main [15] were among the first ethologists to compile and publish a comprehensive review of positive welfare assessment methodology, suggesting that positive welfare should be evaluated on the basis of input (physical resources that are required or valued by an animal) as well as output (positive outcomes such as behavioral responses, cognitive processes and physiological markers indicative of positive affect) [15].

Experimental approaches that might allow researchers to explore animal happiness include physiological, behavioral and cognitive assessment. Yeates and Main [15] noted the utility of analyzing each source of evidence in context; for example, a behavior pattern appearing to be connected to a positive emotional experience can be corroborated by its display in appropriate contexts indicative of positive affect. Behavioral indicators of positive welfare involve activities typically performed when an animal's affective state is positive, such as, behavioral reactions to pleasant taste, vocalizations associated with outcomes expected to entail positive affect, and facial expressions associated with positive experiences [15]. Though the facial expressions of fishes are not well studied, many species of fish are known to be capable of producing distinct vocalizations that could be useful in discerning their general affect and it is reasonable to expect the possibility that fish display a range of physical and behavioral reactions in response to pleasant sensation (including changes in facial conformation).

Later, Balcombe [16] expanded on the evolutionary significance of positive affect, reiterating Fraser and Duncan's [7] claim that pleasure is adaptive as it motivates animals to partake in activities that benefit their fitness and evaluating potential sources of pleasure in animal life. In so doing, Balcombe [16] presented an overview of common behaviors likely to initiate or indicate positive affect, such as playing, engaging in sexual activity, eating and tactile stimulation.

*2.2. Current Thinking on Positive Animal Welfare*

The last decade has seen a marked increase in interest in positive animal welfare and proposed applications facilitating positive affect in animal husbandry guidelines. A number of prominent ethologists have advocated for incorporating positive welfare measures into existing welfare assessment paradigms. For example, Mellor [17] suggests incorporating positive welfare assessment into the Five Freedoms [18], arguing that fulfilling the Five Freedoms is not sufficient to ensure animal welfare as it ignores important psychological aspects of welfare beyond the animal's immediate physical and biological needs. Similarly, Mellor and Beausoleil [19] propose modifying the Five Domains model of animal welfare developed by Mellor and Reid to include a measure of positive welfare in each domain (see [17] for a table of Five Freedoms and updates for positive welfare).

Other scientists have proposed incorporating methods used in human psychology into existing animal welfare assessment paradigms. Mellor [20] presents a neuropsychological explanation of the relationship between emotions and behavior in animals based on a model of action-oriented systems previously described in humans. Webb et al. [21] suggest adopting human quality of life indicators into animal welfare assessment methodology as a framework for quantifying and assessing animal happiness.

Most recently, ethologists have proposed methods of incorporating positive animal welfare indicators into animal husbandry and management guidelines. Lawrence, Newberry and Spinka [22] propose applying positive welfare assessments in order to improve the quality of animal life in pig-keeping, highlighting the overwhelming bias towards negative welfare research thus far. Similarly, Whitham and Wielebnowski [23] and Wolfensohn [24] suggest incorporating positive welfare into standard animal welfare assessment guidelines used by zoos and aquariums (such as the Animal Welfare Assessment Grid).

## 3. Motivation and Environmental Inputs

What types of environments and opportunities are likely to create positive welfare for fishes? In general, animal welfare relies on understanding what the animals want, which provides a starting place for testing ideas regarding what inputs are likely to be needed for ensuring welfare [25]. For example, a seminal study in mink (*Neovison vison*) used behavioral economic principles and methodology to show that mink will push heavily weighted doors to access swimming pools but will not push as hard for access to slightly more (empty) space. Importantly, this program of research also showed that consistent with their motivational preferences, mink experience increased stress when they are prevented from having access to swimming opportunities, but not when they are prevented from having access to slightly more space [26]. Recent studies have used similar experimental designs in fish to show, for example, that goldfish will swim against a current to access plants [27] and that a species of cichlid, Mozambique tilapia (*Oreochromis mossambicus*), will push against a door to access social partners [28]. Before the experiment is designed and the data collected, however, the model of motivation inspiring the study plays an outsized and sometimes hidden role in the type of information that is gained from the research [29].

Thus far, the predominant model of motivation in animal welfare has been an implicit model that is focused on modifying physical parameters within the current systems of use, e.g., agriculture, laboratories, zoos and aquaria. While this informal 'model' has provided invaluable information regarding the elements that need to be changed to mitigate and alleviate potential suffering, it has been less useful for answering questions related to positive welfare. The creation of positive welfare requires a more sophisticated model of motivation, one that is not merely reactionary to currently suboptimal conditions, but instead identifies promising areas of investigation for the goal of creating a full and fulfilling life.

Recent work in human psychology provides one such model of motivation. Higgins [30] has identified three domains to consider in motivation research: (1) value effectiveness, the motivation to have valued outcomes (and avoid aversive ones), (2) control effectiveness, the motivation to be in control or be agentic, and (3) truth effectiveness, the motivation to learn and have opportunities for cognitive engagement. Evidence for these three domains can be found throughout the animal kingdom [12], indicating that this model may also serve as a useful framework for developing hypotheses related to what inputs may promote positive welfare in fish.

### 3.1. Motivation for Value Effectiveness

To date, the most studied domain of motivation in animal welfare is value effectiveness: the motivation to have or to avoid specific material outcomes. Classic animal welfare research that falls into this category has shown, for instance, that hens are highly motivated to access nest boxes before laying [31]. Work related to value effectiveness has been conducted in many species of fish as well. One recent study used a preference index to assess Nile tilapias' (*Oreochromis niloticus*) motivation over time for different tank wall colors [32] and other work has shown that male tilapia's (*O. mossambicus*) substrate preference is conditional on social context [33]. Similarly, other work has shown that Coho salmon (*Oncorhynchus kisutch*) have a strong preference to be in environments with dark backgrounds and, when restricted to lighter backgrounds, are four times more aggressive than when they have access to dark backgrounds [34].

While value effectiveness is often used to identify husbandry features that might cause suffering by their presence (e.g., aversive ambient conditions) or absence (e.g., preferred substrate), it can also stimulate hypotheses for resources that might promote positive affective states. For example, research designed to identify which food sources a fish is most motivated to eat (regardless of nutritional content), could provide ideas for food items to periodically introduce into their diet to create sensory pleasure [16]. Similarly, studies assessing the motivation to access hiding structures would be a first step to understanding what design features may best promote a sense of safety and security in fish.

### 3.2. Motivation for Control Effectiveness

Control effectiveness refers to the desire to manage or control what happens [30]. In other words, beyond the desire to simply have good outcomes (which would be classified as value effectiveness), studies in many different species have indicated that animals are also motivated to be agentic, to be the one who brings the desired outcome about [12,35]. Support for the importance of a distinct motivation for control effectiveness can be found across the animal behavior and welfare literature, including recent work on animal agency [2,11], the proposed relationship between predictability and animal welfare [36], and contrafreeloading studies, some of which date back to the middle of the last century [37]. The common thread throughout all these lines of research is the evidence that animals are interested in taking action, often rejecting opportunities to be passive recipients of desired end-states and benefiting from opportunities have more control over their lives.

While we are not aware of any research directly assessing motivation for control effectiveness in fish, there are several studies that indicate it is a promising domain to consider. For example, Nile tilapia were found to have improved immune function and lower stress levels when they had control over their feeding schedule [38]. This study employed a yoked design in which half the fish were given access to a device that allowed them to trigger the release of food on demand. The other half of the fish received food on a 'yoked' schedule, matched to when the first group activated the food device. This design thus kept the material outcomes and their timing constant across both conditions; the only difference was that one group had control (agency) whereas the other did not. As such, the improved immune function (which, in addition to positive welfare, may have added production value), can be traced to the increased agency in the self-feeding group.

More work in the domain of control effectiveness in fish is clearly needed and the large body of evidence across species of terrestrial animals [11,12] points towards several promising areas for future fish welfare research. In particular, contrafreeloading studies—which assess the degree to which animals will perform unnecessary work (e.g., lever pressing or foraging behaviors) for freely available resources—could investigate whether fish contrafreeload and if they do, determine what sorts of work they enjoy most.

Another methodology that may be useful for examining whether fish are motivated to have control is conditioned place preference (CPP) paradigms, which involve training an individual to associate a neutral environment with a certain type of experience (e.g., receiving a reward). If the animal perceives the experience to be positive, the originally neutral environment will eventually become preferred and if the animal perceives the experience to be negative, the originally neutral environment will eventually become aversive. As CPPs have already been successfully demonstrated in some species of fish including zebrafish (*Danio rerio*) [39] and gilthead sea bream (*Sparus aurata*) [40], CPP studies represent a promising avenue for future research of this kind. If fish are motivated to have control or be agentic, we would expect future research to show that fish develop CPPs for environments in which they experience control opportunities.

### 3.3. Motivation for Truth Effectiveness

The third domain to consider is truth effectiveness, the motivation to learn, explore, and engage in cognitively stimulating activities [30]. The motivation for truth effectiveness is distinct from the motivations for value effectiveness and control effectiveness in that it refers to the motivation for gathering and processing information. In other words, beyond the desire to simply have good outcomes (value effectiveness) and to make good outcomes happen (control effectiveness), studies in many different species have indicated that animals are also motivated to explore, solve problems, and understand patterns [12,35].

Evidence for truth effectiveness motivation in nonhuman animals dates back to classic studies from the middle of the 20th century, with Harlow's [41] work on problem-solving behavior in monkeys and Berlyne's [42] investigations into curiosity in rats. Since then, many other species have been found to want to engage in information processing activities, causing scientists to refer to a fundamental exploratory motive [43] and link problem-solving opportunities to animal welfare [44]. Most recently, there is mounting interest in animal boredom, which is a potentially underappreciated welfare risk and, without appropriate cognitive stimulation, one which many captive animals are likely to suffer [3,4,45].

Existing evidence suggests that at least some species of fish have a motivation for truth effectiveness. Anecdotally, fish are often noted for their curiosity, i.e., the motivation for information—for example, the observation that in experimental settings, some species of fish are known to explore a maze even after it has been solved. There is also some scientific work to support this impression [14]. For example, experimental work has shown that European minnows (*Phoxinus phoxinus*) have a strong propensity to inspect novel objects [46] and when given the opportunity, zebrafish will readily inspect new spaces [47].

Novelty and exploration are not inevitably motivating to animals, however. Most species find forced exposure to novel environments and unfamiliar objects to be aversive, which is why these manipulations are often used to assess negative emotions such as anxiety, stress, and fear. Indeed, one of the most commonly used tests of negative affective state in fish is the novel tank diving paradigm [48]. The key difference between these types of tests and those that can reveal an animal's desire for exploration is whether exposure to the novelty is voluntary or forced. Under forced-choice exploration paradigms, animals often show signs of aversion, whereas under free-choice paradigms, animals often show signs of preference for information gain [43]. Thus, future work aimed at assessing the degree to which fish species may be motivated for learning, exploration, and problem solving (i.e., truth effectiveness), should employ free-choice paradigms whenever possible and ensure that fish have access to a familiar, safe shelter in which they can retreat if a novel experience becomes overwhelming.

Providing fish with environmental enrichment [49] designed to increase complexity while mimicking their natural habitat may be an effective way to offer choice, encourage exploration, and facilitate curiosity. While determining appropriate enrichment strategies is highly dependent on the species' natural history, there are some principles that have been found to be successful in generating positive effects on a variety of species used in research and aquaculture. In general, many species of captive fish benefit from the incorporation of species-appropriate substrate, tank color, structural complexity and sensory stimulation provided by aquatic plants and submerged structures such as logs or rocks. For a comprehensive overview of current environmental enrichment research for captive fishes, see Näslund and Johnsson [49].

## 4. Positive Emotion

The study of emotions in animals has received a growing interest in the past few decades, as evidenced by the emergence of a new discipline referred to as Affective Neuroscience [50] and fish research has been no exception to this trend [1,51]. From an evolutionary perspective, emotions are considered to be adaptive programs intended to direct other physiological responses or directly solve adaptive problems faced by a species over time, defined as intense but short-lived affective responses to an event manifested through physical changes [50]. Emotion is classically described as including a behavioral component, an autonomic component (physiological and endocrine responses) and a subjective component (emotional experience or feeling), and can thus be assessed using both physiological and behavioral parameters [50].

### 4.1. Physiological Indicators of Positive Emotion

Changes in physiological parameters such as heart rate, blood pressure and levels of adrenaline and/or cortisol are often used as indicators of emotions. However, physiological markers should ideally be paired with a behavioral assessment as physiology on its own may indicate sensory arousal but not necessarily emotional valence [51].

The link between rewarding behavior and positive emotion has been established through both behavioral and neuropsychological analysis. Appetitive (seeking/wanting) behaviors or incentive salience have been associated with the release of dopamine from the mesolimbic pathway, a neural pathway found in most vertebrates (including fish) that is responsible for facilitating reinforcement and reward-related motor function learning and plays a role in the subjective perception of pleasure [52], while consummatory (liking) behaviors have been associated with opioids [50]. Social behavior, reproductive behavior, play, self-grooming and other maintenance behaviors all have rewarding properties, releasing endorphins which activate the mesolimbic system, establishing a positive feedback loop and corresponding behavioral expression. Conversely, deprivation of opportunities to perform specific rewarding behaviors produces signs of withdrawal paralleling those observed in deprivation of an addictive drug, further reinforcing the link between rewarding behavior and pleasurable experience [50].

Emotional limbic activity has been shown effect immune function in vertebrates, suggesting that immune activity may prove a promising tool in investigating emotion [50]. In pigs, positive and negative emotional experiences affected immune system parameters (s-IgA levels) antagonistically, paralleling results of experiments on the relationship between emotion and immune function in humans [50]. It should be noted that immune system function is influenced by a plethora of external and internal factors and should therefore only be used to assess emotional valence in in conjunction with other physiological or behavioral markers.

Physiological evidence suggests that fish possess the necessary anatomical and chemical structures to experience emotion [6,51]. It appears that emotions involve brain circuits that have been conserved and elaborated upon throughout vertebrate evolution [53]. The fish telencephalon has homologous limbic and dopaminergic structures involved in emotional processing similar to the mammalian amygdala and hippocampus. Additionally, fish show stress responses similar to mammals, releasing adrenaline and noradrenaline during fear and stress responses, resulting in increased heart rate and ventilation. Fish also produce cortisol in response to emotional stimuli and express transmitter substances including dopamine, serotonin and oxytocin/isotocin associated with reward and mood in humans [54,55]. Finally, multiple studies have shown that fish possess the physiological criteria for pain perception including nociceptors, natural pain-killers (opioids), neural pathways and other brain areas involved in pain perception [56–58].

### 4.2. Behavioral Indicators of Positive Emotion

A number of behaviors, many of which have been documented in fish [14], can signal positive emotional experience in animals, including successful coping, accessing reward, and successful goal-directed behavior. Boissy et al. [50] propose monitoring five behavioral expressions of positive welfare; play behavior, affiliative behavior, self-grooming, vocalizations, and information-gathering. Most research on positive emotion in fish has focused on affiliative and play behaviors, the latter of which is discussed in the section on play.

A potential source of positive emotion in gregarious species is communication and social bonding [59]. As social bonding is a critical element of survival in many species, it follows that pro-social behaviors incorporating fitness benefits may indicate positive emotional states. Pro-social affiliation is likely a good starting place to explore positive emotions in fish as they have various communication modalities [51] and diverse and complex social behaviors.

For example, many fish species have the ability to recognize each other as individuals and some also exhibit parental care [51]. Similarly, some species of cichlids form monogamous pairs and have helpers, constructing a social group of fish of both sexes who share in defending territory and the brood [51]. Studies on zebrafish (*Danio rerio*) have shown that proximity to a conspecific has rewarding properties and that even mere visual contact with a conspecific can be used to reinforce behavior in an associative learning task [60]. Allogrooming represents another group activity that has been linked to positive emotion, such as inducing a relaxed state [50] in fish and other vertebrates. In fish, it has been shown that tactile stimulation can lower stress levels [61] and facilitate future pro-social interactions [62]. Taken as a whole, these lines of research indicate that, under certain circumstances, the presence of a conspecific and pro-social behavior is likely to involve positive emotional responses in some species of fish.

Following this line of reasoning, Franks et al. [63] evaluated a distinct type of shoaling behavior, heightened shoaling, in order to determine its candidacy as a behavioral indicator of positive emotion in zebrafish. Continuous scans of six 30-L tanks of zebrafish housed in semi-natural conditions (sloping gravel substrate, artificial plants, and rocks) revealed multiple instances of this unique and discrete behavior: distinct episodes of abnormally high group cohesion and synchrony, typically lasting for approximately 7 min, but sometimes as long as 30 min. Because these episodes of heightened shoaling were also characterized by low aggression and seemed to occur spontaneously, due to internal group dynamics alone, they share many of the features that are associated with positive emotion in other species [50]. This study underlines the need for a better understanding of species-specific behavior under naturalistic condition in order to successfully evaluate the true behavioral repertoire that may be indicative of positive emotion [7,50].

Communication is another promising area of study, although it should be noted that traditional methods of evaluating social communication in animals may not be applicable to fishes. Rather than relying on anthropocentric behavioral manifestations of pleasure such as facial expressions, we suggest investigating alternate behavioral indicators of positive affect specific to fish, including vocalizations [64], chemical signaling [51], coloration changes, electrical impulses, and bioluminescence. In sum, there is a diversity of potential for future work in identifying behaviors indicative of positive emotions in fish.

### 4.3. Eliciting Positive Emotion in Fish

Boissy et al. [50] propose several practical ways of promoting positive experiences in farmed animals, in both the short and long term: positive anticipation (signaling or predicting a reward in advance), positive contrast (giving a higher reward than expected) and coping and controllability (enabling the animals to cope with or to control the situation). These interventions can all be applied to and tested in fish, suggesting many fruitful ideas for how positive emotions might be elicited and studied in fish.

Anticipation of a positive or rewarding event induces a state of positive affect. Investigations of what animals find positively reinforcing and of the animal's behavioral expression during anticipation or expectation of the rewarding result provide a basis for assessing positive emotional states in animals [50]. Such work (some of it dating back decades) has already been conducted in several species of fish. A recent example comes from work by Nilsson [65], who used conditioning and trained groups of cod (*Gadus morhua*) to associate a light signal with a positive up-coming event, i.e., the delivery of food. The fish showed anticipatory behavior and gathered in the feeding area after they had been presented with the light signal.

Positive contrast, on the other hand, describes the experience of receiving an outcome that is better than expected. In a typical positive contrast experiment, an animal is first trained to expect a reinforcer of a certain size. The test phase occurs when the expected reward is replaced by an even stronger or larger reinforcer (e.g., more food or better-quality food). If the animal changes its behavior so that its response is faster/more vigorous than that of a control group given the larger reinforcer from the start, a successive positive contrast (SPC) has occurred [50]. Interestingly, an early study by Bottjer, Scobie and Wallace [66] found behavior in goldfish (*Carassius auratus*) that is consistent with a positive contrast effect. In this study, goldfish in the positive contrast condition responded at a higher rate to the conditioned stimulus (a light cue) than fish in the control condition, despite the fact that the reinforcement level (i.e., food amount) was the same for both groups. Given the cognitive sophistication found in many species of fish [5], it is likely that other species of fish may experience positive contrast (i.e., are able to track the relative value of a reward rather than it's absolute value), which suggests that positive contrast may be a promising area for future research on positive emotion in fish.

Finally, scholars have suggested that successful coping and exerting control over the environment can also increase positive welfare [12]. While facing challenges inevitably involves some degree of stress, only chronic stress brought about by the persistent inability to cope causes long-term distress and learned helplessness [50]. In contrast, short-term social and physical challenges can be perceived as positive engagement experiences when the net result is a learned coping behavior and/or rewarding goal. In other words, if there is a high probability of success, challenges can actually be a source of positive welfare. Studies applying learning procedures with positive reinforcement, for example, are likely to reflect the behavior of fish experiencing positive emotion in response to a cognitive challenge.

## 5. Play

Play has fascinated scholars for millennia and yet it remains a highly contentious area of study for many reasons, not the least of which is that play behavior can be hard to define and identify [7,13]. Nonetheless, part of the root of the scientific fascination with play stems in part from the notion that under certain conditions, play may be associated with positive emotions and signal improved welfare. Indeed, some scholars have identified it as one of the most promising indicators of positive emotions in domestic animals in part because it can be measured noninvasively and can be easily recognized in some mammals [7,13,50]. Identifying and measuring play behavior in non-mammalian species is, however, much more challenging due to its variability between species and mutable characteristics [67]. Moreover, the relationship between play and welfare and especially positive emotions is complex and controversial [13,68]. Thus connections between play and positive welfare require more work before the implications of observing different types of play at different levels in different species held under different conditions are known. With due consideration of these qualifications, however, play remains an intriguing avenue for future research on positive welfare in fish for reasons we outline below.

### 5.1. Identifying and Assessing Play in Fishes

First, a large body of anecdotal evidence and some empirical work suggests that play exists in many species of fish [14,67]. How might we assess play behavior in fish? Ethologist Gordon Burghardt [14] presents a working method for identifying play behavior based on five criteria. While aspects of these criteria can be somewhat subjective, they represent a useful starting point for flagging behaviors that may be good candidates for future research. Burghardt's core criteria are as follows:

1. Play behaviors are typically not fully functional and are often directed toward stimuli that do not contribute to immediate survival.
2. Play is autotelic (self-rewarding) and is often characterized as something that animals seem to engage in voluntarily. Play behaviors appear to be spontaneous, intrinsically motivated, rewarding, reinforcing or pleasurable.
3. Play behaviors may resemble 'serious' performance of ethotypic behavior but differ in structure and/or timing to adult behavior. Play behavior is "incomplete, exaggerated, awkward, or precocious" and "involves behavior patterns with modified form, sequencing, or targeting" [14].
4. Play is performed repeatedly, but not stereotypically during at least one developmental period in life" [14].
5. Play is initiated when the animal is in a relaxed state, i.e., when there are no immediate threats to the animal's fitness and the animal is not strongly motivated to perform other behaviors.

Burghardt [14] argues that fish possess the necessary behavioral and cognitive pre-requisites that support play behavior in other animals, including exploration and curiosity. The existing evidence for play in fish are discussed below along with the main types of play.

Social play involves two or more animals usually of similar age and size [13,69] and facilitates integration into groups and social skills [70,71]. As such, social play is more commonly observed in young animals than adults, though it is known to occur at lower frequencies in adults of some species [14]. Burghardt [14] reports anecdotal accounts suggestive of social play in several cichlid species. For example, Burghardt [14] describes an instance in which a socially-housed white spotted cichlid (*Tropheus duboisi*) was observed repeatedly approaching a cichlid species from Lake Malawi (*Labeotropheus sp.*), performing a lateral display and eliciting a brief chase, a potential example of teasing play [14]. Similarly, young cichlid (sp) are often observed nibbling and rubbing against their parent's flanks, which may be play behavior as it does not appear to serve any adaptive function [14]. Controlled and systematic observation of these types of fish social behavior, especially between neonatal cichlids and their parents, would be useful for determining whether they are consistent with our understanding of play.

Object play is directed at inanimate objects and can involve one or more participants [14]. It is typically observed when animals manipulate objects that appear to provide no obvious survival benefit [14] and is thought to be related to the development of motor skills [69]. Observations of object play have been observed across a wide range of fish species. For example, many species of captive fish have been reported repeatedly manipulating objects within their environment for no immediately obvious function [14] and anecdotal evidence suggests some predatory species such as great white sharks (*Carcharodon carcharias*) may hunt for fun when prey is not needed for nourishment. Most recently, an empirical study systematically observed the object manipulation behavior of white spotted cichlids (*Tropheus dubois*) and found that their behavior was consistent with all the criteria for play behavior [67]. The provision of sensory-stimulating objects may be a good starting place for further empirical work on object play in fish.

Locomotor play involves jumping, running and performing other motor activities in a sudden, persistent and frenetic manner [14,69,71]. It is thought that locomotor play serves to benefit the animal physiologically by promoting muscle growth, strengthening bones and increasing cardiopulmonary capability [13]. Perhaps the most well-documented example of locomotor play in fish is leapfrogging, in which individuals repeatedly launch themselves over objects floating in the water such as plants, leaves, sticks and other animals. Burghardt [14] notes that leapfrogging has been reported in needlefish, *Hemiramphidae* (halfbeaks), *Clupeidae* (herring and shad) and *Atherinidae* (silversides) and outlines several hypotheses attempting to explain the origins and purposes of leapfrogging, including the idea that leapfrogging may be an instinctive means of ectoparasite control or to elicit a pleasurable tactile sensation. Most recently, Fagen [72] has argued that jumping behavior is likely a form of locomotor behavior in Atlantic salmon (*Salmo salar*).

*5.2. The Play–Welfare Relationship*

Play is often used to assess affective states in both human and nonhuman animals [14,68] and is gaining interest in applied ethology as a welfare indicator [50,69]. The first clue that play behavior is important to animal welfare is its absence in poor welfare situations. Play behavior often disappears when animals are under fitness challenges but re-emerge as they recover [13]. As such, the absence of play, particularly in young animals, is considered an indication of poor welfare.

Importantly, however, play behaviors also change in timing or kind in conjunction with a decrease in welfare. Research in children suggests that, beyond the total duration of play, poor welfare may be reflected in qualitative aspects such as the display of solitary over social play and the degree to which play bouts are continuous vs. fragmented. Play research aimed at understanding the play–welfare relationship should, therefore, attend to the quality (type and fragmentation) and timing in addition to its quantity [68]. Future work on fish play could begin by mapping out the range of potential play behaviors and its qualitative characteristics across different conditions, which is a necessary starting point before strong inferences to welfare are drawn.

Overall, however, play is most common in situations when the animal is well fed and not under any threat to fitness [68]. This positive association between play behavior and need fulfillment provides the basic foundation for the link between play behavior and animal welfare [13,50]. Fraser and Duncan [7] argue that certain behaviors are indicative of internal states; according to their theory of animal welfare, play occurs when an animal is in an opportunity situation in which their immediate physical needs are met and the behavior results in a net positive impact to their fitness (i.e., the pleasure derived from performing the activity eclipses the fitness cost of performing that activity).

Play can be used both a behavioral indicator of good welfare but also as tool to promote it [13]. Several studies have documented the contagious nature of play amongst conspecifics, suggesting that one playful individual can trigger a cascade of positive affect, resulting in substantial welfare benefits to social animals. Social learning, which involves an element of contagious behavior, has been documented extensively in teleost species [73,74], suggesting that future research on fish social interaction may yield important results to welfare. For example, Held and Spinka [13] suggested stimulating social play behavior in captive animals by providing the opportunity for positive social communication, a scenario that could be applied to captive fish by housing individuals in social groups of appropriate size and composition.

Future research with fish could leverage these intriguing ideas and follow up on the anecdotal observations outlined by Burghardt [14]. Importantly, however, such research requires housing fish in environmental and social conditions conducive to a relaxed state. Furthermore, the current relative paucity of documented play behavior in fish may not indicate that fish do not play but rather that they are too uncomfortable in the typical housing we provide for them to engage in play. The first steps towards studying play behavior and by extension the relationship between play and welfare in fish is therefore to provide animals with social and environmental conditions conducive to a relaxed state in which their needs are met.

## 6. Conclusions

Drawing from existing literature on welfare paradigms in both human and nonhuman animals, we identified potential physiological and behavioral signifiers likely involved with positive welfare in fish and suggest future research that could support the ways in which caretakers can promote positive welfare in captive fish including the provision of species-appropriate housing (e.g., ambient color) and social conditions as well as opportunities for control (e.g., self-feeders) and cognitive engagement (e.g., visual stimulation via port holes and/or novel objects). Incorporating indicators of positive affect into fish welfare assessment strategies will be particularly important given the growing number of captive fish for whose wellbeing humans are primarily responsible. In particular, generating scientific knowledge on ways of identifying and monitoring species-specific indications of positive affect is an essential step forward in understanding fish welfare.

In closing, it is worth clarifying that we are not advocating for positive welfare as a distinct field of research. Separating positive welfare from negative welfare is likely an artificiality, creating the impression of welfare as a static contrast between immediate pleasure and immediate suffering. In psychology, it has been argued that there is no positive psychology or negative psychology, there is only psychology [75]; similarly, rather than categorizing welfare into positive vs. negative types, it may be more useful to have the simple goal of understanding welfare and making sure that that understanding is complete. Thus, the aim of this review was not to present a new and separate area of study but rather to highlight a class of experiences and emotions that are currently being neglected and issue the call for more research in these areas.

**Author Contributions:** Individual contributions include: Conceptualization, B.F. & I.F.-C.; investigation, B.F. & I.F.-C.; writing—original draft preparation, B.F. & I.F.-C.; writing—review and editing, B.F. & I.F.-C.; supervision, B.F.; funding acquisition, B.F. & I.F.-C.

**Funding:** This research was funded in part by a generous donation to the Fish and Marine Animal Welfare Fund at New York University. IFC received support from the Animal Law and Policy (ALP) Program at UCLA.

**Conflicts of Interest:** The authors declare no conflict of interest. The funders had no role in the design of the study; in the collection, analyses, or interpretation of data; in the writing of the manuscript, or in the decision to publish the results.

## References

1. Cerqueira, M.; Millot, S.; Castanheira, M.F.; Félix, A.S.; Silva, T.; Oliveira, G.A.; Oliveira, C.C.; Martins, C.I.M.; Oliveira, R.F. Cognitive Appraisal of Environmental Stimuli Induces Emotion-like States in Fish. *Sci. Rep.* **2017**, *7*, 13181. [CrossRef]
2. Špinka, M. Animal Agency, Animal Awareness and Animal Welfare. *Anim. Welf.* **2019**, *28*, 11–20. [CrossRef]
3. Meagher, R. Is Boredom an Animal Welfare Concern? *Anim. Welf.* **2019**, *28*, 21–32. [CrossRef]
4. Burn, C.C. Bestial Boredom: A Biological Perspective on Animal Boredom and Suggestions for Its Scientific Investigation. *Anim. Behav.* **2017**, *130*, 141–151. [CrossRef]
5. Brown, C. Fish Intelligence, Sentience and Ethics. *Anim. Cogn.* **2015**, *18*, 1–17. [CrossRef] [PubMed]

6.  Braithwaite, V.A.; Huntingford, F.; van den Bos, R. Variation in Emotion and Cognition Among Fishes. *J. Agric. Environ. Ethics* **2013**, *26*, 7–23. [CrossRef]
7.  Fraser, D.; Duncan, I.J.H. "Pleasures", "pains" and Animal Welfare: Toward a Natural History of Affect. *Anim. Welf.* **1998**, *7*, 383–396.
8.  Mellor, D.J. Moving beyond the "Five Freedoms" by Updating the "Five Provisions" and Introducing Aligned "Animal Welfare Aims". *Animals* **2016**, *6*, 59. [CrossRef]
9.  Seligman, M.E.P.; Csikszentmihalyi, M. Positive Psychology—An Introduction. *Am. Psychol.* **2000**, *55*, 5–14. [CrossRef]
10. Fredrickson, B.L. Positive Emotions Broaden and Build. In *Advances in Experimental Social Psychology*; Devine, P.G., Plant, A., Eds.; Academic Press: Burlington, MA, USA, 2013; pp. 1–53.
11. Špinka, M.; Wemelsfelder, F. Environmental Challenge and Animal Agency. In *Animal Welfare*; Appleby, M.C., Ed.; CAB International: Wallingford, UK, 2018; pp. 39–55.
12. Franks, B.; Higgins, E.T. Effectiveness in Humans and Other Animals: A Common Basis for Well-Being and Welfare. *Adv. Exp. Soc. Psychol.* **2012**, *46*, 285–346. [CrossRef]
13. Held, S.D.E.; Špinka, M. Animal Play and Animal Welfare. *Anim. Behav.* **2011**, *81*, 891–899. [CrossRef]
14. Burghardt, G.M. The Origins of Vertebrate Play: Fish That Leap, Juggle, and Tease. In *The Genesis of Animal Play: Testing the Limits*; MIT Press: Cambridge, MA, USA, 2005.
15. Yeates, J.W.; Main, D.C.J. Assessment of Positive Welfare: A Review. *Vet. J.* **2008**, *175*, 293–300. [CrossRef]
16. Balcombe, J. Animal Pleasure and Its Moral Significance. *Appl. Anim. Behav. Sci.* **2009**, *118*, 208–216. [CrossRef]
17. Mellor, D.J. Updating Animal Welfare Thinking: Moving beyond the "Five Freedoms"; towards "A Life Worth Living". *Animals* **2016**, *6*, 21. [CrossRef]
18. Brambell, F. *Report of the Technical Committee to Enquire into the Welfare of Animals Kept under Intensive Livestock Husbandry Systems*; Her Majesty's Stationery Office: London, UK, 1965.
19. Mellor, D.; Beausoleil, N. Extending the "Five Domains" Model for Animal Welfare Assessment to Incorporate Positive Welfare States. *Anim. Welf.* **2015**, *24*, 241–253. [CrossRef]
20. Mellor, D. Animal Emotions, Behaviour and the Promotion of Positive Welfare States. *N. Z. Vet. J.* **2012**, *60*, 1–8. [CrossRef]
21. Webb, L.E.; Veenhoven, R.; Harfeld, J.L.; Jensen, M.B. What Is Animal Happiness? *Ann. N. Y. Acad. Sci.* **2019**, *1438*, 62–76. [CrossRef]
22. Lawrence, A.B.; Newberry, R.C.; Špinka, M. Positive Welfare: What Does It Add to the Debate over Pig Welfare? In *Advances in Pig Welfare*; Elsevier, Inc.: Amsterdam, The Netherlands, 2018. [CrossRef]
23. Whitham, J.C.; Wielebnowski, N. New Directions for Zoo Animal Welfare Science. *Appl. Anim. Behav. Sci.* **2013**, *147*, 247–260. [CrossRef]
24. Wolfensohn, S.; Shotton, J.; Bowley, H.; Davies, S.; Thompson, S.; Justice, W.; Wolfensohn, S.; Shotton, J.; Bowley, H.; Davies, S.; et al. Assessment of Welfare in Zoo Animals: Towards Optimum Quality of Life. *Animals* **2018**, *8*, 110. [CrossRef]
25. Dawkins, M.S. From an Animal's Point of View: Motivation, Fitness, and Animal Welfare. *Behav. Brain Sci.* **1990**, *13*, 1–9. [CrossRef]
26. Mason, G.J.; Cooper, J.; Clarebrough, C. Frustrations of Fur-Farmed Mink. *Nature* **2001**, *410*, 35–36. [CrossRef] [PubMed]
27. Sullivan, M.; Lawrence, C.; Blache, D. Why Did the Fish Cross the Tank? Objectively Measuring the Value of Enrichment for Captive Fish. *Appl. Anim. Behav. Sci.* **2016**, *174*, 181–188. [CrossRef]
28. Galhardo, L.; Almeida, O.; Oliveira, R.F. Measuring Motivation in a Cichlid Fish: An Adaptation of the Push-Door Paradigm. *Appl. Anim. Behav. Sci.* **2011**, *130*, 60–70. [CrossRef]
29. Franks, B. What Do Animals Want? *Anim. Welf.* **2019**, *28*, 1–10. [CrossRef]
30. Higgins, E.T. *Beyond Pleasure and Pain: How Motivation Works*; Oxford University Press: New York, NY, USA, 2012.
31. Cooper, J.J.; Appleby, M.C. Motivational Aspects of Individual Variation in Response to Nestboxes by Laying Hens. *Anim. Behav.* **1997**, *54*, 1245–1253. [CrossRef] [PubMed]
32. Maia, C.M.; Volpato, G.L. Preference Index Supported by Motivation Tests in Nile Tilapia. *PLoS ONE* **2017**, *12*, e0175821. [CrossRef] [PubMed]
33. Galhardo, L.; Almeida, O.; Oliveira, R.F. Preference for the Presence of Substrate in Male Cichlid Fish: Effects of Social Dominance and Context. *Appl. Anim. Behav. Sci.* **2009**, *120*, 224–230. [CrossRef]

34. Gaffney, L.P.; Franks, B.; Weary, D.M.; von Keyserlingk, M.A.G. Coho Salmon (*Oncorhynchus kisutch*) Prefer and Are Less Aggressive in Darker Environments. *PLoS ONE* **2016**, *11*, e0151325. [CrossRef]

35. Higgins, E.T.; Cornwell, J.F.M.; Franks, B. "Happiness" and "The Good Life" as Motives Working Together Effectively. In *Advances in Motivation Science*; Academic Press: Cambridge, MA, USA, 2014; Volume 1, pp. 135–179.

36. Bassett, L.; Buchanan-Smith, H.M. Effects of Predictability on the Welfare of Captive Animals. *Appl. Anim. Behav. Sci.* **2007**, *102*, 223–245. [CrossRef]

37. Osborne, S.R. Free food (contrafreeloading) phenomenon—Review and analysis. *Anim. Learn. Behav.* **1977**, *5*, 221–235. [CrossRef]

38. Endo, M.; Kumahara, C.; Yoshida, T.; Tabata, M. Reduced Stress and Increased Immune Responses in Nile Tilapia Kept under Self-Feeding Conditions. *Fish. Sci.* **2002**, *68*, 253–257. [CrossRef]

39. Wong, D.; von Keyserlingk, M.A.; Richards, J.G.; Weary, D.M. Conditioned Place Avoidance of Zebrafish (*Danio Rerio*) to Three Chemicals Used for Euthanasia and Anaesthesia. *PLoS ONE* **2014**, *9*, e88030. [CrossRef] [PubMed]

40. Millot, S.; Cerqueira, M.; Castanheira, M.F.; Øverli, Ø.; Martins, C.I.M.; Oliveira, R.F. Use of Conditioned Place Preference/Avoidance Tests to Assess Affective States in Fish. *Appl. Anim. Behav. Sci.* **2014**, *154*, 104–111. [CrossRef]

41. Harlow, H.F. Learing and satiation of response in intrinsically motivated comples puzzle performance by monkeys. *J. Comp. Physiol. Psychol.* **1950**, *43*, 289–294. [CrossRef] [PubMed]

42. Berlyne, D.E. Curiosity and Exploration. *Science* **1966**, *153*, 25–33. [CrossRef] [PubMed]

43. Hughes, R. Intrinsic Exploration in Animals: Motives and Measurement. *Behav. Processes* **1997**, *41*, 213–226. [CrossRef]

44. Meehan, C.L.; Mench, J.A. The Challenge of Challenge: Can Problem Solving Opportunities Enhance Animal Welfare? *Appl. Anim. Behav. Sci.* **2007**, *102*, 246–261. [CrossRef]

45. Franks, B. Cognition as a Cause, Consequence, and Component of Welfare. In *Advances in Agricultural Animal Welfare*; Woodhead Publishing: Duxford, UK, 2018.

46. Murphy, K.E.; Pitcher, T.J. Individual Behavioural Strategies Associated with Predator Inspection in Minnow Shoals. *Ethology* **1991**, *88*, 307–319. [CrossRef]

47. Graham, C.; von Keyserlingk, M.A.G.; Franks, B. Free-Choice Exploration Increases Affiliative Behaviour in Zebrafish. *Appl. Anim. Behav. Sci.* **2018**, *203*, 103–110. [CrossRef]

48. Egan, R.J.; Bergner, C.L.; Hart, P.C.; Cachat, J.M.; Canavello, P.R.; Elegante, M.F.; Elkhayat, S.I.; Bartels, B.K.; Tien, A.K.; Tien, D.H.; et al. Understanding Behavioral and Physiological Phenotypes of Stress and Anxiety in Zebrafish. *Behav. Brain Res.* **2009**, *205*, 38–44. [CrossRef]

49. Näslund, J.; Johnsson, J.I. Environmental Enrichment for Fish in Captive Environments: Effects of Physical Structures and Substrates. *Fish Fish.* **2016**, *17*, 1–30. [CrossRef]

50. Boissy, A.; Manteuffel, G.; Jensen, M.B.; Moe, R.O.; Spruijt, B.; Keeling, L.J.; Winckler, C.; Forkman, B.; Dimitrov, I.; Langbein, J.; et al. Assessment of Positive Emotions in Animals to Improve Their Welfare. *Physiol. Behav.* **2007**, *92*, 375–397. [CrossRef]

51. Kittilsen, S. Functional Aspects of Emotions in Fish. *Behav. Processes* **2013**, *100*, 153–159. [CrossRef] [PubMed]

52. O'Connell, L.A.; Hofmann, H.A. The Vertebrate Mesolimbic Reward System and Social Behavior Network: A Comparative Synthesis. *J. Comp. Neurol.* **2011**, *519*, 3599–3639. [CrossRef] [PubMed]

53. Chandroo, K.; Duncan, I.J.; Moccia, R. Can Fish Suffer? Perspectives on Sentience, Pain, Fear and Stress. *Appl. Anim. Behav. Sci.* **2004**, *86*, 225–250. [CrossRef]

54. Winberg, S.; Nilsson, G.E. Roles of Brain Monoamine Neurotransmitters in Agonistic Behaviour and Stress Reactions, with Particular Reference to Fish. *Comp. Biochem. Physiol. Part C Pharmacol. Toxicol. Endocrinol.* **1993**, *106*, 597–614. [CrossRef]

55. Thompson, R.R.; Walton, J.C. Peptide Effects on Social Behavior: Effects of Vasotocin and Isotocin on Social Approach Behavior in Male Goldfish (*Carassius Auratus*). *Behav. Neurosci.* **2004**, *118*, 620–626. [CrossRef]

56. Braithwaite, V. *Do Fish Feel Pain?* Oxford University Press: New York, NY, USA, 2010.

57. Sneddon, L.U. Pain in Aquatic Animals. *J. Exp. Biol.* **2015**, *218 Pt 7*, 967–976. [CrossRef]

58. Sneddon, L.U. The Evidence for Pain in Fish: The Use of Morphine as an Analgesic. *Appl. Anim. Behav. Sci.* **2003**, *83*, 153–162. [CrossRef]

59. Rolls, E.T. On the Brain and Emotion. *Behav. Brain Sci.* **2000**, *23*, 219–228. [CrossRef]

60. Al-Imari, L.; Gerlai, R. Sight of Conspecifics as Reward in Associative Learning in Zebrafish (*Danio Rerio*). *Behav. Brain Res.* **2008**, *189*, 216–219. [CrossRef]

61. Soares, M.C.; Oliveira, R.F.; Ros, A.F.H.; Grutter, A.S.; Bshary, R. Tactile Stimulation Lowers Stress in Fish. *Nat. Commun.* **2011**, *2*, 534. [CrossRef] [PubMed]
62. Bshary, R.; Würth, M. Cleaner Fish Labroides Dimidiatus Manipulate Client Reef Fish by Providing Tactile Stimulation. *Proc. R. Soc. Lond. Ser. B Biol. Sci.* **2001**, *268*, 1495–1501. [CrossRef]
63. Franks, B.; Graham, C.; von Keyserlingk, M.A.G. Is Heightened-Shoaling a Good Candidate for Positive Emotional Behavior in Zebrafish? *Animals* **2018**, *8*, 152. [CrossRef] [PubMed]
64. Ghazali, S. Fish Vocalisation: Understanding Its Biological Role from Temporal and Spatial Characteristics. Ph.D. Thesis, Auckland University, Auckland, New Zealand, 2011.
65. Nilsson, J.; Kristiansen, T.S.; Fosseidengen, J.E.; Fernö, A.; van den Bos, R. Learning in Cod (*Gadus Morhua*): Long Trace Interval Retention. *Anim. Cogn.* **2008**, *11*, 215–222. [CrossRef] [PubMed]
66. Bottjer, S.W.; Scobie, S.R.; Wallace, J. Positive behavioral contrast, autoshaping, and omission responding in the goldfish (*Carassius auratus*). *Anim. Learn. Behav.* **1977**, *5*, 336–342. [CrossRef]
67. Burghardt, G.M.; Dinets, V.; Murphy, J.B. Highly repetitive object play in a cichlid fish (*Tropheus duboisi*). *Ethology* **2015**, *121*, 38–44. [CrossRef]
68. Ahloy-Dallaire, J.; Espinosa, J.; Mason, G. Play and Optimal Welfare: Does Play Indicate the Presence of Positive Affective States? *Behav. Processes* **2018**, *156*, 3–15. [CrossRef]
69. Oliveira, A.F.S.; Rossi, A.O.; Silva, L.F.R.; Lau, M.C.; Barreto, R.E. Play Behaviour in Nonhuman Animals and the Animal Welfare Issue. *J. Ethol.* **2010**, *28*, 1–5. [CrossRef]
70. Spinka, M.; Newberry, R.C.; Bekoff, M. Mammalian play: Training for the unexpected. *Q. Rev. Biol.* **2001**, *76*, 141–168. [CrossRef]
71. Bekoff, M. Social Play Behavior. *Bioscience* **1984**, *34*, 228–233. [CrossRef]
72. Fagen, R.M. Salmonid Jumping and Playing: Potential Cultural and Welfare Implications. *Animals* **2017**, *7*, 42. [CrossRef]
73. Brown, C.; Laland, K. Social Learning and Life Skills Training for Hatchery Reared Fish. *J. Fish Biol.* **2001**, *59*, 471–493. [CrossRef]
74. Brown, C.; Laland, K.N. Social Learning in Fishes: A Review. *Fish Fish.* **2003**, *4*, 280–288. [CrossRef]
75. McNulty, J.K.; Fincham, F.D. Beyond Positive Psychology? Toward a Contextual View of Psychological Processes and Well-Being. *Am. Psychol.* **2012**, *67*, 101–110. [CrossRef]

![fishes logo]

*Review*

# A Global Assessment of Welfare in Farmed Fishes: The FishEthoBase

João Luis Saraiva [1,*], Pablo Arechavala-Lopez [1,2], Maria Filipa Castanheira [1], Jenny Volstorf [1,3] and Billo Heinzpeter Studer [1,3]

1 Fish Ethology and Welfare Group, CCMAR, 8005-139 Faro, Portugal; pablo@fair-fish.net (P.A.-L.); maria-filipa@fair-fish.net (M.F.C.); jenny@fair-fish.net (J.V.); international@fair-fish.net (B.H.S.)
2 IMEDEA-UIB, 07190 Esporles, Illes Balears, Spain
3 Fair-Fish International Association, CH-8355 Aadorf, Switzerland
* Correspondence: joao@fair-fish.net; Tel.: +351-289-800-051

Received: 15 February 2019; Accepted: 7 May 2019; Published: 16 May 2019

✓ check for updates

**Abstract:** Fish welfare is an essential issue that needs to be tackled by the aquaculture industry. In order to address it, studies have been limited to a small number of species and the information is generally scattered. In order to have a consistent overview of the welfare of farmed fishes, we present the FishEthoBase, an open-access database that ultimately aims to provide information on the welfare of all fish species currently farmed worldwide. Presently with 41 species, this database is directed to all stakeholders in the field and targets not only to bridge the gaps between them but also to provide scientific information to improve the welfare of fish. The current text explains the database and presents an analysis of the welfare scores of 41 species, suggesting that (i) the general welfare state of farmed fishes is poor, (ii) there is some potential for improvement and (iii) this potential is related to research on species' needs, but (iv) there are many remaining knowledge gaps and (v) current fish farming technologies do not seem to fully address welfare issues. The existence of a framework, such as the FishEthoBase, is proposed as fundamental to the design of strategies that improve the welfare of farmed fish.

**Keywords:** fish welfare; ethology; FishEthoBase; risk analysis; welfare scores; welfare criteria; framework

## 1. Introduction

Aquaculture is presently the main source of sea food products for human consumption, with approximately 80 million tonnes of harvested fish in 2016 [1]. However, unlike the relatively low number of terrestrial species farmed for food (26 in total), the fish production relies on 362 species [2–4]. This is an astonishing number that fluctuates as new taxa are added each year while others are abandoned [2–4], although it is worth noting that the 20 most common species account for 84.2% of total fish production [5]. In addition, and again in contrast with the amount of information on the biology of terrestrial farm animals, there are severe knowledge gaps regarding many (if not most) farmed aquatic species. The knowledge that does exist is focused primarily on production traits rather than welfare. In fact, research has pushed the physiological limits of many fish species in growth, fertility and size, as a consequence of (or resulting in) highly artificial conditions [6]. This creates an obvious issue regarding welfare: Fish are sentient beings [7–9] and each species has evolved for millennia in natural contexts, developing adaptations, behaviours and coping mechanisms which are relevant for those contexts [10]. The artificial conditions of captivity, particularly in industrial aquaculture, pose a whole new category of stimuli, for which the animals are seldom equipped to deal with: Space restraints, unnatural aggregations, barren environments, handling and other frequent artificial stressors, etc. [11]. Artificial selection may not necessarily be an answer to the issue, because (1) the domestication of fish is very recent and (2) selected or 'domesticated' strains may be far from their welfare optima as a consequence of their domestication processes [3].

Natural behaviour is pivotal for the welfare assessment of captive animals. Behaviour is the first and foremost indicator of the biological state of an animal, and behavioural observations are the best tool to understand not only the physiological state of the individual but also its mental state [3,12–14]. Therefore, the knowledge on the ethology of farmed fish species is a conditio sine qua non for the correct evaluation of their welfare. This evaluation must rely on a clear operational definition of welfare and on robust indicators that are able to measure unambiguously the variables they are assumed to be measuring. Optimal welfare indicators should take into account not only the health of the fish (i.e., a function-based approach to welfare) [15] but also reflect the animals' emotional state (i.e., a feelings-based approach) [16] as well as their natural needs (i.e., a nature-based approach) [11,17]. In addition, assessing cognitive biases [18] and positive affective states [19] is now an important tool to improve the welfare of farmed fish. A suite of operational welfare indicators has been developed extensively for salmon, for example [20,21], while a broader framework of behavioural indicators is also available for fishes in general [14].

A definition of welfare that complies with the criteria previously addressed is: "The state of the individual as it copes with the environment" [22,23]. This definition of welfare has several implications:

> "1) Welfare is a characteristic of an animal, not something that is given to it; 2) Welfare will vary from very poor to very good, i.e., the individual may be in a poor state at one end of the welfare continuum or in a good state at the other, 3) Welfare can be measured objectively and independently of moral considerations; 4) Measures of failure to cope and measures of how difficult it is for an animal to cope both give information about how poor the welfare is; 5) Knowledge of the preferences of an animal often gives valuable information about what conditions are likely to result in good welfare, but direct measurements of the state of the animal must also be used in attempts to assess welfare and improve it; and 6) Animals may use a variety of methods when trying to cope. There are several consequences of failure to cope, so any one of a variety of measures can indicate that welfare is poor, and the fact that one measure, such as growth, is normal does not mean that welfare is good".

([22], page 4168)

It is well known that domesticated land animals still display behaviours that reflect adaptation to their evolutionary environments, despite not being necessary in their rearing contexts [24]. Considering that this phenomenon occurs in species that have been under domestication for millennia, it is more than likely to occur also in fish, whose domestication process is far more recent: Centuries in the case of carp, tilapia and trout, or decades as in the case of most of the cultured fish recently [3]. The behavioural patterns of fishes observed in the wild are, therefore, the gold standard to assess their welfare state. Although phenotypic plasticity may surely play a role in the adaptation of a fish species to artificial rearing [25–28] it is nevertheless imperative to understand the ethology or natural behavioural repertoire of farmed animals in order to evaluate their welfare conditions and implement measures to mitigate the effects of life in captivity.

In this paper we will:

- Describe in detail the FishEthoBase project, an open-access database on fish ethology and welfare.
- Use these data to assess the general welfare state of farmed fishes presently and use the scoring scheme in the database (FishEthoScore) to provide an outlook on the potential of fish species to be farmed in good welfare.

## 2. The FishEthoBase Project

The database aims to provide the basis for the global assessment of the welfare of farmed fish. The public portal www.fishethobase.net is a platform where the scientific knowledge regarding the ethology of ultimately all farmed aquatic species is reviewed, scrutinised, organised and summarised in order to answer relevant questions concerning welfare. The database, therefore, aims to bridge the gap between the scientific community and the aquaculture industry. The data are organised into two main approaches:

(1)  Full profiles, divided into Findings, where an extensive review on the biology of the species in the wild and in captivity is assessed through bibliographical reviews. Recommendations, where proposals for rearing in captivity under good welfare conditions are made on the basis of the review, and a Summary where this knowledge is condensed.
(2)  Short profiles, where a rapid assessment of the welfare state of each species is performed through literature-based answers to 10 core criteria, pointing at main problems and possible solutions, and providing the base for the numerical assessment of welfare-the FishEthoScore.

The fundamental idea underlying the FishEthoBase is to devise a series of criteria that can be applied to all farmed species, and therefore create a framework that allows the comparison of species in terms of welfare. By asking the same questions about all species, not only can we compare their welfare state, but also assess and rank their potential to be farmed in good welfare conditions. The details of the methodology and rationale are explained below.

*2.1. Full Profiles—Findings*

The findings section on the database aims to be a systematic and exhaustive literature review of the essential knowledge on the behavioural biology of each farmed species, with the purpose of understanding in depth how the farming conditions affect the fish. All entries are referenced, identified in the text with a number, and the reference list is detailed by citing order at the bottom of each species' profile. The following criteria, containing detailed sub-criteria, are addressed:

- Ethograms
- Distribution
- Natural co-existence
- Substrate and/or shelter
- Food, foraging, hunting, feeding
- Photoperiod
- Water parameters
- Swimming
- Growth
- Reproduction
- Senses
- Communication
- Social behaviour
- Cognitive abilities
- Personality, coping styles
- Emotion-like states
- Self-concept, self-recognition
- Reactions to husbandry

*2.2. Full Profiles—Recommendations*

The data presented in the findings section is used to create a series of recommendations, which provide essential tools for the humane rearing of the species reviewed. These are not meant to be equivalent in detail to a farming manual but rather present how systems should be designed to accommodate the fishes' needs. For example, it is likely that the recommendations discourage from certain practices that studies have indicated to be detrimental to welfare or suggest alternative methods that have been demonstrated to improve the welfare state.

*2.3. Summary*

All the previous data is condensed in a shorter and less technical text to deliver the main information to a broader audience.

*2.4. Short Profiles*

The Short Profiles Section of the database aims to be a sharp assessment tool of the general welfare state of a farmed species. To achieve this, a set of 10 core criteria was chosen to portray the variables in fish farming most likely to affect the welfare of fish (see Section 2.4.1 for a description of why and how the criteria were selected). Each criterion is divided into the general life stages of the animal, which generally correspond to rearing stages in the farming environment: Eggs, larvae (hatchery), juveniles (nursery), adults (grow-out) and spawners (broodstock). The entries for each life stage then refer to the knowledge in nature (i.e., in the "Wild") and under aquaculture conditions (i.e., in "Farm"), ideally discriminating the various farming methods when literature is available. For each criterion, the existing knowledge on the biology in the wild is, therefore, overlapped with the existing knowledge on farming conditions, which will implicitly allow drawing conclusions on the welfare conditions of the species regarding that criterion. Furthermore, this overlap is the basis for the scoring of the FishEthoScore, an index that summarises the likelihood of good welfare under common rearing conditions, the potential for rearing under good welfare in high-standard conditions and the certainty of our findings (explained in detail in Section 2.4.2).

The formalisation logic of the entries is summarised in Figure 1.

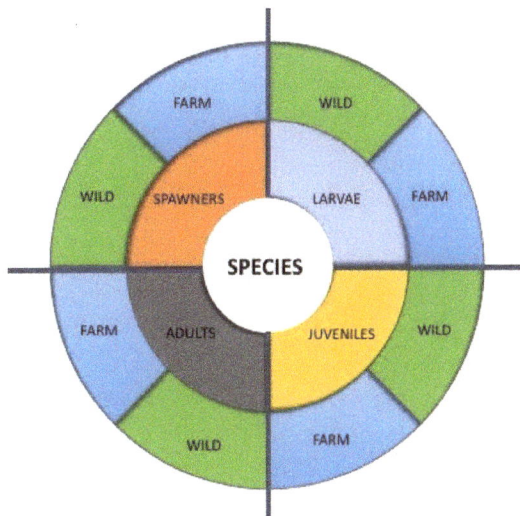

**Figure 1.** Logic and organization of the entries in the Short Profiles. Each species is divided into four main life stages (larvae, juveniles, adults and spawners), corresponding to four production stages (hatchery, nursery, grow-out and broodstock). For each of these, the entries provide information from the wild and under farming conditions (see text for details).

All entries in the database are referenced, identified in the text with a number, and the reference list is detailed by citing order at the bottom of each species' profile. When there is no reliable information on any of the sections described above, a standard sentence 'no data found yet' is entered. When findings are contradictory or insufficient, the entry becomes 'further research needed … ' to highlight the existence of knowledge gaps.

### 2.4.1. Selected Criteria for the Short Profiles

The basic rationale for the short profiles is that the catalogue of questions, or criteria, designed to achieve a rapid evaluation of the welfare state of a farmed species should be as short and sharp as possible. We arbitrarily set the cut off line at 10 critical questions, which should (i) depict the major limitations imposed to the lives of fish under farming conditions and, therefore, directly impact their welfare (ii) be able to be applied to all farmed species. These criteria were designed to take into account not only the multidimensional nature of welfare (mental, physiological and natural [29–31]) but also common conceptual guidelines towards animal welfare in practice—namely the five freedoms [29] and the allostatic model [32,33].

In general, the constraints imposed on fish in any farming method are: Restricted space, manipulation and handling, low complexity of the environment, unnatural aggregation of individuals, artificial feed and feeding regime, and slaughter. The selection of the 10 core criteria, therefore, reflects these impositions (Table 1).

**Table 1.** Description and rationale of criteria used in the short profiles of the FishEthoBase www. fishethobase.net.

| Criterion | Type of Constraint | Possible Effects |
|---|---|---|
| 1. Home range | Spatial | Disrupted swimming behaviour, impaired movement. |
| 2. Depth range | Spatial | Disrupted swimming behaviour, impaired movement. |
| 3. Migration and habitat change | Spatial | Disrupted swimming behaviour, migration drivers not met (feeding, reproduction, habitat choice, etc.). |
| 4. Free reproduction | Physiological, Behavioural | Impaired mating, courting and spawning behaviours, inbreeding, disrupted sexual selection. |
| 5. Aggregation | Social | Impaired communication and swimming patterns, disrupted social networks, territoriality and shoaling configurations, increased overall cost of high social stress. |
| 6. Aggression | Behavioural, social | Increased fighting and injuries, increased overall cost of high social stress. |
| 7. Substrate and shelter | Environmental, ecological | Altered swimming and/or feeding patterns, reduced opportunities for escape, disrupted flight response, increased overall cost of high social stress. |
| 8. Handling | Physiological, Mental | Infection, injury, anxiety, chronic stress, pain, death. |
| 9. Malformations | Physiological | Impaired mobility, feeding, breathing or other aspects of biology. |
| 10. Slaughter | Death | Extreme pain and suffering. |

### 1. Home Range

By definition, farming fish implies holding individuals in a confined area. The dimensions of such horizontal space restriction vary greatly according to the rearing method, the species and the life stage. In some species and life stages these dimensions may potentially match the natural behaviour of the species (e.g., *C. idella* [34], *O. niloticus* [35,36], *S. aurata* [37,38], *C. gariepinus* [39,40]) while in most cases this does not happen and may, therefore, have a negative effect in fish welfare [41].

## 2. Depth Range

Similarly to the home range, all farming methods require a depth limitation to avoid the escape of fish. In species living in shallow waters, their natural depth range is matched by some farming methods and solutions. However, there is a huge variation in physical and environmental components depending on the depth of the water column [42,43]. Therefore, limiting the natural vertical movements to depths that are feasible for aquaculture systems may directly impair fish welfare. Furthermore, shallow depths may also interfere with welfare through higher exposure to anthropogenic disturbances [44,45], inadequate exposure to light [44] and ultraviolet radiation [46] or forced contact with surface-dwelling disease agents [47].

## 3. Migration and Habitat Change

Spatial restraints become especially relevant if they impede migration. The negative effects may be mitigated if farmers provide the animals with the resources they are migrating for, e.g., access to mates, target feed or season-appropriate environmental conditions (for example, *C. gariepinus* moving shallower to spawn which could be simulated by draining water [48], *S. salar* moving from freshwater to the sea at smolt stage [49] may be mimicked by transferring the fish to saltwater). In most cases, however, the welfare of species that perform seasonal movements due to spawning, feeding, environmental or other needs may be strongly affected by captivity because often the motivation to migrate is not yet known, or whether providing the resource negates the desire or need to migrate, or even it is impossible to meet under farming conditions [50].

## 4. Reproduction

In broodstocks, natural reproduction (i.e., without any type of direct manipulation of the animals) is generally accepted as a sign of good welfare: Spontaneous spawning only occurs if the stressors in the captive environment are not strong enough to inhibit the neuroendocrine reproductive pathways [51]. Therefore, the investment in breeding will only be worthwhile for an individual if conditions are suitable. In many broodstocks, however, reproduction is induced through hormonal stimulation and/or spawning is achieved through stripping or other invasive procedures to streamline production [51]. The welfare of the fish may thus be impaired directly by the procedure and indirectly by the associated handling [52] (see also 8. Handling below).

## 5. Aggregation

To streamline fish farming, the animals need not only to be confined but also to be aggregated in order to minimise spatial requirements and maximise catch per effort. There are many ways to calculate and define the number of fish per unit of space or volume and time, and the concept itself is complex [52,53]. However, the artificial manipulation of inter-individual distances during extended periods of the lifetime of fishes may surely represent a major impairment of their welfare: Aggregation below biologically relevant densities may be an issue in naturally shoaling or schooling species, while high density in solitary species may cause stress, infection, behavioural disorders or mortality [11]. The responses to rearing densities are nevertheless specific to species and life stage [11,52,54].

## 6. Aggression

The rearing of aggressive species, or the facilitation of aggressive behaviour in farmed species due to confinement, density or farming method, is not advisable for obvious reasons: Injuries, stress, decrease in production and overall hindering of welfare [11]. For example, chronic subordination in hierarchical species (that are often prevalent in the aquaculture industry, such as salmonids and sparids) leads to serious welfare problems in low-ranking individuals [55–59]. Aggressive interactions may also indirectly impair other aspects in fish life, such as nutrition [60], growth [61] or the immune system [62] for example.

7. Substrate and Shelter Needs

Fish farming hygiene standards and routine activities (cleaning, feeding, grading, harvesting, etc.) generally require simple or barren environments: For example, raceways and tanks are advised to have smooth surfaces to ensure water flow and prevent the accumulation of detritus [63,64]. Sea cages are usually barren for the same reason. While pelagic species might not need substrate or shelter, benthic species rely on substrate features for shelter, nesting or other uses that are an essential part of their natural environment (e.g., *S. aurata*, *O. niloticus* or *S. solea*). When such species are farmed under standard methods, their natural requirements are not met, and their welfare may be impaired. This also applies to auxiliary species, such as lumpfish used to de-lice salmon in sea cages [65]: This species is naturally benthic throughout most of its life [66] and would require a substrate to meet its natural needs.

8. Handling

Manipulation and handling are the most direct acute and chronic stressors that occur in aquaculture. These including routine actions, such as cleaning and culling, regular events, such as grading, sexing, transport, vaccination or other treatments, and high impact procedures, such as crowding, harvesting and slaughtering. The effects of manipulation and handling on welfare are well documented: Physical, physiological and mental stress, injury, infection and death [11,50,52]. These routines are the most invasive actions imposed on farmed aquatic animals and the ones that most highly disturb and impair their welfare [67]. However, these aspects may readily be improved, as they depend solely on engineering solutions.

9. Malformations

Malformations can and do occur naturally in fishes. However, many abnormalities found in farmed species arise, are evoked by or occur in higher frequency due to the farming conditions themselves: Collisions due to spatial restrictions or early stage incorrect rearing temperatures may be responsible for skeletal deformities, common diseases or absence of optimum water quality may cause morphological problems, nutritional or feeding deficiencies or may easily delay or impair growth and cause cataracts, for example [68,69]. While taking into account that the lack of predation and lower competition found in the aquaculture environment allow longer survival of deformed fish (at least theoretically), these individuals have, nevertheless, their welfare impaired by deformations prompted by human action [70].

10. Stunning and Slaughter

Humane slaughter is a key feature for the good welfare of farmed fish. However, most farmed species are slaughtered commercially through asphyxia in ice [71], despite the fact that there are presently many solutions that ensure either immediate and painless death or render the fish unconscious through effective stunning [71–78] These solutions, however, are not applied as consistently as expected, which results in a vast majority of aquaculture production being slaughtered in poor conditions that impose excruciating and unnecessary suffering to individuals. In addition, humane slaughtering of fish results in better overall product quality [79].

Supplemental Criterion (A): Domestication Level

The artificial selection processes that led to domestication in terrestrial animals are in effect in fish. Theoretically, the selection of strains that show better adaptation to the rearing environment would possibly benefit the welfare of the animals. However (1) the concept of 'better adaptation' is often anthropomorphic and based on production indicators, not welfare [3] and (2) The artificial selection of fish is very recent, and despite its intensity, it is not clear whether we can truly refer to fish as domesticated [3,4]. Therefore, although it is undeniable that human-induced selection occurs, its effect on welfare is often ambiguous and scoring domestication level as positive or negative in welfare terms becomes equivocal (see Section 2.4.2).

Supplemental Criterion (B): Forage Fish in the Feed

The final question of the short profiles deals not only with the welfare of the focal species but with the ecological and welfare influence of forage fisheries on harvested animals. The undeniable impact of this activity should not be taken lightly [80] but the welfare effects of fishmeal and fish oil replacement with alternative sources are not yet fully determined: There are reports of health and development problems of farmed individuals using fish meal and fish oil replacements [81]. Therefore, scoring forage fishery and replacement with sustainable sources as positive or negative in welfare terms is confusing (see Section 2.4.2).

The data compiled, summarised and critically reviewed in these criteria is the backbone of the database. It should allow the user to make an informed evaluation of the current and potential welfare state of the species and show where there are knowledge gaps that deserve further research, regardless of the additional scoring we provide and that is described in the following section.

2.4.2. Scoring Framework: The FishEthoScores

In addition to providing data on each of the criteria, the information is also summarised in welfare scores: The FishEthoScores are extracted from the entries, and the scoring framework is based on a common risk analysis background [82] to condense the existing information and determine:

- The **Likelihood** that the fish experience good welfare under the lowest standard farming conditions found in the literature regarding that specific criterion. The possible scores are Low, High, Unclear or No Findings;
- The **Potential** for the species to experience good welfare in the highest standard rearing conditions regarding that specific criterion (or the expected improvements in the near future). The possible scores are Low, Medium, High, Unclear or No Findings.
- The **Certainty** of our assessment of the criterion, i.e., a measurement of the general quality, quantity and clarity of the data available for the species. The possible scores are Low, Medium, High or No Findings.

The questions, answers and the scoring framework are logically designed so that affirmative responses have a positive valence, while negative responses have a negative valence (Table 2).

**Table 2.** Questions regarding the 10 selected criteria for the Short Profiles in the FishEthoBase www.fishethoase.net.

| Criterion | Question(s) | Possible Answers |
|-----------|-------------|------------------|
| 1 | Are minimal farming conditions likely to provide the home range of the species? | Likelihood is **High**.<br>Likelihood is **Low**.<br>Data are **Unclear**. |
| | Is there potential for improvement? | Potential is **High**.<br>Potential is Medium.<br>Potential is Low.<br>Data are **Unclear**.<br>There are **No findings** to support scoring. |
| | How certain are these findings? | Certainty is **High**.<br>Certainty is **Medium**.<br>Certainty is **Low**.<br>There are **No findings** to support scoring. |

*Fishes* **2019**, *4*, 30

**Table 2.** *Cont.*

| Criterion | Question(s) | Possible Answers |
|---|---|---|
| 2 | Are minimal farming conditions likely to provide the depth range of the species? | Likelihood is **High**.<br>Likelihood is **Low**.<br>Data are **Unclear**. |
| | Is there potential for improvement? | Potential is **High**.<br>Potential is Medium.<br>Potential is **Low**.<br>Data are **Unclear**.<br>There are **No findings** to support scoring. |
| | How certain are these findings? | Certainty is **High**.<br>Certainty is **Medium**.<br>Certainty is **Low**.<br>There are **No findings** to support scoring. |
| 3 | Are minimal farming conditions compatible with the migrating or habitat-changing behaviour of the species? | Likelihood is **High**.<br>Likelihood is **Low**.<br>Data are **Unclear**. |
| | Is there potential for improvement? | Potential is **High**.<br>Potential is **Medium**.<br>Potential is **Low**.<br>Data are **Unclear**.<br>There are **No findings** to support scoring. |
| | How certain are these findings? | Certainty is **High**.<br>Certainty is **Medium**.<br>Certainty is **Low**.<br>There are **No findings** to support scoring. |
| 4 | Is the species likely to reproduce in captivity without manipulation? | Likelihood is **High**.<br>Likelihood is **Low**.<br>Data are **Unclear**. |
| | Is there potential to allow for it under farming conditions? | Potential is **High**.<br>Potential is **Medium**.<br>Potential is **Low**.<br>Data are **Unclear**.<br>There are **No findings** to support scoring. |
| | How certain are these findings? | Certainty is **High**.<br>Certainty is **Medium**.<br>Certainty is **Low**.<br>There are **No findings** to support scoring. |
| 5 | Is the aggregation imposed by minimal farming conditions likely to be compatible with the natural behaviour of the species? | Likelihood is **High**.<br>Likelihood is **Low**.<br>Data are **Unclear**. |
| | Is there potential to allow for it under farming conditions? | Potential is **High**.<br>Potential is **Medium**.<br>Potential is **Low**.<br>Data are **Unclear**.<br>There are **No findings** to support scoring. |
| | How certain are these findings? | Certainty is **High**.<br>Certainty is **Medium**.<br>Certainty is **Low**.<br>There are **No findings** to support scoring. |

**Table 2.** *Cont.*

| Criterion | Question(s) | Possible Answers |
| --- | --- | --- |
| 6 | Is the species likely to be non-aggressive and non-territorial? | Likelihood is **High**.<br>Likelihood is **Low**.<br>Data are **Unclear**. |
| | Is there potential for improvement? | Potential is **High**.<br>Potential is **Medium**.<br>Potential is **Low**.<br>Data are **Unclear**.<br>There are **No findings** to support scoring. |
| | How certain are these findings? | Certainty is **High**.<br>Certainty is **Medium**.<br>Certainty is **Low**.<br>There are **No findings** to support scoring. |
| 7 | Are minimal farming conditions likely to match the natural substrate and shelter needs of the species? | Likelihood is **High**.<br>Likelihood is **Low**.<br>Data are **Unclear**. |
| | Is there potential for improvement? | Potential is High.<br>Potential is **Medium**.<br>Potential is **Low**.<br>Data are Unclear.<br>There are **No findings** to support scoring. |
| | How certain are these findings? | Certainty is **High**.<br>Certainty is **Medium**.<br>Certainty is **Low**.<br>There are **No findings** to support scoring. |
| 8 | Are minimal farming conditions (handling, confinement etc.) likely not to stress the individuals of the species? | Likelihood is **High**.<br>Likelihood is **Low**.<br>Data are **Unclear**. |
| | Is there potential for improvement? | Potential is High.<br>Potential is **Medium**.<br>Potential is **Low**.<br>Data are Unclear.<br>There are **No findings** to support scoring. |
| | How certain are these findings? | Certainty is **High**.<br>Certainty is **Medium**.<br>Certainty is **Low**.<br>There are **No findings** to support scoring. |
| 9 | Are malformations of this species likely to be rare under farming conditions? | Likelihood is **High**.<br>Likelihood is **Low**.<br>Data are **Unclear**. |
| | Is there potential for improvement? | Potential is **High**.<br>Potential is **Medium**.<br>Potential is **Low**.<br>Data are **Unclear**.<br>There are **No findings** to support scoring. |
| | How certain are these findings? | Certainty is **High**.<br>Certainty is **Medium**.<br>Certainty is **Low**.<br>There are **No findings** to support scoring. |

Table 2. *Cont.*

| Criterion | Question(s) | Possible Answers |
|---|---|---|
| 10 | Is a humane slaughter protocol likely to be applied under minimal farming conditions? | Likelihood is **High**. Likelihood is **Low**. Data are **Unclear**. |
| | Is there potential for improvement? | Potential is **High**. Potential is **Medium**. Potential is **Low**. Data are **Unclear**. There are **No findings** to support scoring. |
| | How certain are these findings? | Certainty is **High**. Certainty is **Medium**. Certainty is **Low**. There are **No findings** to support scoring. |

The FishEthoScores for each species follow a conservative strategy and are determined through the simple sum of the high scores in each dimension:

Likelihood FishEthoScore: The sum of 'High' scores throughout the criteria, varies between 0 and 10.
Potential FishEthoScore: The sum of 'High' scores throughout the criteria, varies between 0 and 10.
Certainty FishEthoScore: The sum of 'High' scores throughout the criteria, varies between 0 and 10.

The FishEthoScores are, therefore, a conservative quantitative summary of a qualitative assessment. The strategy to exclusively use the sum of 'High' scores and avoid attributing intermediate values to other possible scores ("Medium", "Low", "Unclear" or "No findings") is deliberate. In that sense, the FishEthoScores may be a simplification (which any index inherently is) but we opt to lose detailed information so that the calculation of the score is consistent and straightforward. In any case, the scoring information is available both in all the text entries and summarised in a table the top section of each short profile.

Taken together, the three FishEthoScores aim to be representative of the overall welfare state for each species under farming conditions, i.e., high FishEthoScores represent good welfare whereas low FishEthoScores represent poor welfare.

The dimensions (Likelihood, Potential and Certainty) are independent of each other, therefore allowing comparisons and rankings among species and among criteria. The FishEthoBase is highly dynamic, is continuously growing and ever-evolving as new species are added and new papers are published. For this analysis, we will use the data publicly available on 31 October 2018 (41 species).

## 3. Assessment of Welfare in Farmed Fish Species

A complete table of all the species analysed with their corresponding FishEthoScores is provided as Supplementary Material.

Using the FishEthoScores, we analysed the meaningful correlations between them. We used the scores on Likelihood, Potential, Certainty and Improvement Capacity (measured as Potential minus Likelihood as a proxy to gauge how far a species is from its best possible welfare conditions) in a Spearman correlation matrix, in order to answer questions, such as (but not limited to):

- Are farmed fish species experiencing good welfare?
- Is there potential for farmed species to experience good welfare?
- Are fish species far from their best possible welfare conditions? Does the existing knowledge influence the current or prospective welfare state of farmed fish?
- Are fish species which currently experience better welfare the ones who show the greatest potential to be farmed in best conditions?

All tests were two-tailed with α = 0.05. Correlations and significance tests were performed using Microsoft Excel.

Regarding the low-standard farming conditions, the average *Likelihood* score is 0.44 ± 0.02 (mean ± SE) (in a maximum of 10). The maximum value found is four in the Kingfish *Seriola lalandi* followed by three in the Nile tilapia *Oreochromis niloticus.*

Considering the high-standard practices found in the literature, the average *Potential* is 1.37 ± 0.04 (in a maximum of 10), with a maximum of eight in the Nile tilapia followed by six in the African catfish *Clarias gariepinus.*

The current knowledge regarding the 10 criteria is also sub-optimal, as *Certainty* has an average of 1.93 ± 0.04 with a maximum of six in the Nile tilapia followed by five in African catfish and in the European sea bass *Dicentrarchus labrax.*

There seems to be little *Improvement Capacity*, as on average the species are 0.93 points away (where zero is the complete fulfillment) to achieve their full welfare potential. The case where there is more capacity to improve is the African catfish, where six points separate the best from the worst conditions.

Conversely, the domestication levels of farmed fish according to [83] are high, with an average of 3.90 ± 0.02 (in a maximum of five) and many cases where species are fully domesticated (i.e., there are already selective breeding programs, focused on different traits, such as growth, yield, flesh quality, etc.) [83,84].

The correlation matrix between these variables is summarised in Table 3. We found significant correlations between Likelihood and Potential. The latter is also correlated with Certainty and Improvement Capacity. No other significant correlations are found.

**Table 3.** Correlation matrix for the selected variables. Values are Spearman ρ. Significant correlations are marked with *.

|  | Likelihood | Potential | Certainty | Domestication | Improvement Capacity |
|---|---|---|---|---|---|
| Likelihood | 1 |  |  |  |  |
| Potential | 0.60 * | 1 |  |  |  |
| Certainty | 0.21 | 0.56 * | 1 |  |  |
| Domestication | 0.02 | 0.11 | 0.14 | 1 |  |
| Improvement Capacity | 0.08 | 0.80 * | 0.49 | 0.18 | 1 |

## 4. Discussion

The FishEthoBase is the first (and, to our knowledge, the only) exhaustive database concerning the welfare of farmed fishes. Despite the limitations inherent to any database, it nevertheless provides a broad overview of the variability of farmed fish species, their specificities concerning welfare and the current knowledge regarding the main welfare aspects that may be affected by farming practices. However, perhaps even more importantly, it provides an overview of the empty spaces: Not only does it point out the research gaps but also the mismatch between scientific advances and industry practices. In many cases this mismatch is due to practices implemented in fish farms being ahead of what is known and published in academia. In others, it is the other way around. Therefore, one of the major aims of FishEthoBase is to help clear the opacity between the two fields and favour the communication between them.

According to the review of the literature, the low-standard farming conditions generally fail to respond to the basic welfare needs of farmed fish. Although slightly better, the high-standard conditions found in the literature nevertheless also fail to meet the basic welfare requirements of these species. Even considering that the approach of the FishEthoBase is very broad and conservative, that the industry has advances which take time to reach the literature (and vice-versa) and that any kind scoring is limited to some degree by subjective bias, the scenario seems to be bleak. Specifically concerning the Likelihood that a species is presently being reared in good welfare according to the lowest standards found in the literature, no species reaches a positive mark (i.e., >5). Regarding the

Potential to be reared in good welfare, only two species (Nile tilapia and African catfish) out of the 41 have a positive mark. A possible conclusion is that, according to the criteria and evaluation detailed above and based on the scientific literature available, these are the two species whose biology seems to be most appropriate to cope with captive conditions, which nevertheless need to be of the highest standard. All the other 39 in the database fail to pass this test, which may be due to the incapacity of rearing systems to meet the welfare needs of the species at some point of its life cycle, or to the biology of the species not being suitable for farming, regardless of human appetites and industry inclinations. The notion that some species cope better with farming conditions than others is not new (e.g., see [85]) and it is based on the long and extremely diverse evolutionary history of fish. It should, therefore, be possible to exploit biological traits of fish species in order to achieve viable production while maintaining good welfare and imposing minimal stress. On the other hand, some species may not be suitable to cope with aquaculture (at least using the present methods or in a foreseeable future), and therefore, their farming should be either revised or discouraged. Domestication or selection of phenotypes ("strains") that present better indicators could be an answer. However, it should be noted that the process of domestication is very recent in fish [4,83] and may introduce variability in the fish response to rearing that should be dealt with caution [3,85].

The positive correlation between Likelihood and Potential indicates that the species who seem to cope better with its current farming conditions are the ones who have the greatest potential to experience welfare in high-standard farms. It also means that the potential of a species to be reared under good welfare is dependent on its current welfare state. This may be related to the biology of these species, but it is possible that better solutions are available (or under development) for these species because both academia and the industry realise that good welfare 'pays off' also in terms of production: If fish experiencing good welfare perform well, further improving their welfare will also improve their performance.

One of the key results from this analysis is the strong correlation between Certainty and Potential. This suggests that knowledge and research are key to improve the conditions of farmed fish, which contributes to better production results [86,87]. Welfare has also been shown to be a driver when consumers choose their animal products in some markets [88,89], while in others this driver is present only in some segments of the population [90]. The knowledge provided by the FishEthoBase may, therefore, not only be a key to identify research gaps in the welfare of farmed species, but also paves the way for consumer information and future welfare certification schemes that are science-based, reliable, efficient and verifiable.

One result that may be a cause for concern is the very low Improvement Capacity. This means that, in general, there is not much that can be done to improve the welfare of farmed species. This can be explained either because of the species' biology, specific welfare needs throughout the life cycle or at certain stages, or limitations in farming methods and technology. Rather not surprisingly, the Improvement Capacity is also related to the Potential, which is self-explanatory.

Interestingly, the high domestication level in farmed fishes does not seem to improve their welfare. This is probably because the recent domestication process of fish has been focusing more on production traits than welfare [3], which is also suggested by the low scores on Certainty. In fact, many species have a large body of research regarding feeding, growth or spawning (e.g., tilapia, salmon, trout, sea bream, sea bass) yet the knowledge on welfare traits and indicators is rudimentary, as demonstrated by the low Certainty score. In Europe, recent efforts have been undertaken to address these knowledge gaps. One example is the COST action 'Welfare of fish in European aquaculture', that produced a high number of reports for these species, e.g., [71,73,74,76,77]. Another example is the growing number of papers on welfare of Atlantic salmon [20,54,91–94], that highlights the rising importance of welfare for the farming of this species. Nonetheless, it is still a drop in an empty ocean of knowledge (or lack thereof) for the remaining more than 300 species farmed worldwide.

## 5. Conclusions

In this discussion, we analysed the general trends from the FishEthoScores, and concluded that (i) the general welfare state of farmed fish species is bad, (ii) there is some potential for fish to be reared in better welfare conditions, (iii) this potential is generally related to research on species' needs but (iv) there is much to be done in order to fill in the knowledge gaps and (v) current or prospective fish farming technologies do not seem to allow much improvement. However, we believe we have barely scratched the surface for data exploration. The database is open access and the concept behind it is that the users should explore the portal and its contents to its fullest potential, either in generic terms as we have done here, or in a species-specific manner. Nevertheless, we reiterate the idea that the real essence and value are the data compiled, summarised and critically reviewed for each species. The FishEthoScore is crude yet useful quantification tool to assess the general welfare state of the species—though prone to subjective interpretation. The knowledge underpinning the scores, however, is indeed objective and should allow the user to make an informed evaluation of welfare state of the species, identify the knowledge gaps and eventually design solutions to improve the lives of farmed fish. As new knowledge is gained, the team behind the FishEthoBase continues to update the database. Furthermore, the concept of this project is to provide an open platform for researchers, industry members and other stakeholders to share their expertise and critically examine the data available.

**Supplementary Materials:** The following are available online at http://www.mdpi.com/2410-3888/4/2/30/s1, Table S1: FishEthoScores for 41 farmed fish species available in the open access database FishEthoBase www.fishethobase. net.

**Author Contributions:** Database concept: B.H.S., J.V., J.L.S., P.A.-L., M.F.C.; formal analysis of data, J.L.S.; investigation, B.H.S., J.V., J.L.S., P.A.-L., M.F.C.; data curation, B.H.S., J.V., J.L.S., P.A.-L., M.F.C.; writing—original draft preparation, J.L.S.; review and editing, B.H.S., J.V., J.L.S., P.A.-L., M.F.C.; project administration, B.H.S., J.L.S., J.V.; funding acquisition, B.H.S., J.L.S.

**Funding:** This work was funded by Open Philanthropy Project (San Francisco, USA), Swiss Federal Food Safety and Veterinary Office (Bern, Switzerland), Stiftung Dreiklang (Basel, Switzerland), Haldimann-Stiftung (Aarau, Switzerland) and other private donations. This study received Portuguese national funds from FCT—Foundation for Science and Technology through project UID/Multi/04326/2019.

**Acknowledgments:** We acknowledge many contributions from stakeholders that continuously allow the growing and improvement of the database.

**Conflicts of Interest:** The authors declare no conflict of interest.

## References

1. FAO. *The State of World Fisheries and Aquaculture 2016. Contributing to Food Security and Nutrition for All*; FAO: Rome, Italy, 2016; p. 200.
2. FAO. Domestic Animal Diversity Information System (DAD-IS). Available online: http://www.fao.org/dad-is/data/en/ (accessed on 15 February 2019).
3. Saraiva, J.L.; Castanheira, M.F.; Arechavala-Lopez, P.; Volstorf, J.; Studer, B.H. Domestication and Welfare in Farmed Fish. In *Animal Domestication*; IntechOpen: London, UK, 2018.
4. Teletchea, F.; Teletchea, F. Domestication and Genetics: What a Comparison between Land and Aquatic Species Can Bring? In *Evolutionary Biology: Biodiversification from Genotype to Phenotype*; Springer: Cham, Switzerland, 2015; pp. 389–401.
5. *Meeting the Sustainable Development Goals*; FAO (Ed.) The State of World Fisheries and Aquaculture: Rome, Italy, 2018; ISBN 978-92-5-130562-1.
6. Huntingford, F.A. Implications of domestication and rearing conditions for the behaviour of cultivated fishes. *J. Fish Biol.* **2004**, *65*, 122–142. [CrossRef]
7. Brown, C. Fish intelligence, sentience and ethics. *Anim. Cogn.* **2014**, *18*, 1–17. [CrossRef] [PubMed]
8. Sneddon, L.U. The bold and the shy: Individual differences in rainbow trout. *J. Fish Biol.* **2003**, *62*, 971–975. [CrossRef]
9. Yue, S.; Moccia, R.D.; Duncan, I.J.H. Investigating fear in domestic rainbow trout, Oncorhynchus mykiss, using an avoidance learning task. *Appl. Anim. Behav. Sci.* **2004**, *87*, 343–354. [CrossRef]

10. Helfman, G.; Collette, B.B.; Facey, D.E.; Bowen, B.W. *The Diversity of Fishes: Biology, Evolution, and Ecology*; John Wiley & Sons: Hoboken, NJ, USA, 2009; ISBN 978-1-4443-1190-7.
11. Ashley, P.J. Fish welfare: Current issues in aquaculture. *Appl. Anim. Behav. Sci.* **2007**, *104*, 199–235. [CrossRef]
12. Cerqueira, M.; Millot, S.; Castanheira, M.F.; Félix, A.S.; Silva, T.; Oliveira, G.A.; Oliveira, C.C.; Martins, C.I.M.; Oliveira, R.F. Cognitive appraisal of environmental stimuli induces emotion-like states in fish. *Sci. Rep.* **2017**, *7*, 13181. [CrossRef]
13. Dawkins, M.S. The Science of Animal Suffering. *Ethology* **2008**, *114*, 937–945. [CrossRef]
14. Martins, C.I.M.; Galhardo, L.; Noble, C.; Damsgård, B.; Spedicato, M.T.; Zupa, W.; Beauchaud, M.; Kulczykowska, E.; Massabuau, J.-C.; Carter, T.; et al. Behavioural indicators of welfare in farmed fish. *Fish Physiol. Biochem.* **2012**, *38*, 17–41. [CrossRef] [PubMed]
15. Segner, H.; Sundh, H.; Buchmann, K.; Douxfils, J.; Sundell, K.S.; Mathieu, C.; Ruane, N.; Jutfelt, F.; Toften, H.; Vaughan, L. Health of farmed fish: Its relation to fish welfare and its utility as welfare indicator. *Fish Physiol. Biochem.* **2012**, *38*, 85–105. [CrossRef] [PubMed]
16. Duncan, I.J. A concept of welfare based on feelings. In *The Well-Being of Farm Animals: Challenges and Solutions*; Blackwell Publishing: Ames, IA, USA, 2004; pp. 85–101.
17. Sneddon, L.U. Fish behaviour and welfare. *Appl. Anim. Behav. Sci.* **2007**, *104*, 173–175. [CrossRef]
18. Mendl, M.; Burman, O.H.P.; Parker, R.M.A.; Paul, E.S. Cognitive bias as an indicator of animal emotion and welfare: Emerging evidence and underlying mechanisms. *Appl. Anim. Behav. Sci.* **2009**, *118*, 161–181. [CrossRef]
19. Boissy, A.; Manteuffel, G.; Jensen, M.B.; Moe, R.O.; Spruijt, B.; Keeling, L.J.; Winckler, C.; Forkman, B.; Dimitrov, I.; Langbein, J.; et al. Assessment of positive emotions in animals to improve their welfare. *Physiol. Behav.* **2007**, *92*, 375–397. [CrossRef]
20. Noble, E.C.; Gismervik, K.; Iversen, M.H.; Kolarevic, J.; Nilsson, J.; Stien, L.H.; Turnbull, J.F. *Welfare Indicators for Farmed Atlantic Salmon—Tools for Assessing Fish Welfare*; Nofima: Tromsø, Norway, 2018.
21. Stien, L.H.; Bracke, M.B.; Folkedal, O.; Nilsson, J.; Oppedal, F.; Torgersen, T.; Kittilsen, S.; Midtlyng, P.J.; Vindas, M.A.; Øverli, Ø.; et al. Salmon Welfare Index Model (SWIM 1.0): A semantic model for overall welfare assessment of caged Atlantic salmon: Review of the selected welfare indicators and model presentation. *Rev. Aquac.* **2013**, *5*, 33–57. [CrossRef]
22. Broom, D.M. Animal welfare: Concepts and measurement. *J. Anim. Sci.* **1991**, *69*, 4167–4175. [CrossRef] [PubMed]
23. Broom, D.M. Indicators of poor welfare. *Br. Vet. J.* **1986**, *142*, 524–526. [CrossRef]
24. Bracke, M.B.M.; Hopster, H. Assessing the Importance of Natural Behavior for Animal Welfare. *J. Agric. Environ. Ethics* **2006**, *19*, 77–89. [CrossRef]
25. Dingemanse, N.J.; Kazem, A.J.N.; Réale, D.; Wright, J. Behavioural reaction norms: Animal personality meets individual plasticity. *Trends Ecol. Evol.* **2010**, *25*, 81–89. [CrossRef]
26. Duponchelle, F.; Legendre, M. Rapid phenotypic changes of reproductive traits in response to experimental modifications of spatial structure in Nile tilapia, Oreochromis niloticus. *Aquat. Living Resour.* **2001**, *14*, 145–152. [CrossRef]
27. Hutchings, J.A. Adaptive phenotypic plasticity in brook trout, Salvelinus fontinalis, life histories. *Écoscience* **1996**, *3*, 25–32. [CrossRef]
28. Lorenzen, K.; Beveridge, M.C.M.; Mangel, M. Cultured fish: Integrative biology and management of domestication and interactions with wild fish. *Biol. Rev.* **2012**, *87*, 639–660. [CrossRef]
29. Brambell, F.W.R. *Report of the Technical Committee to Enquire into the Welfare of Animals kept under Intensive Livestock Husbandry Systems*; Farm Animal Welfare Council: London, UK, 1965.
30. Fraser, D. Animal ethics and animal welfare science: Bridging the two cultures. *Appl. Anim. Behav. Sci.* **1999**, *65*, 171–189. [CrossRef]
31. Fraser, D.; Weary, D.; Pajor, E.; Milligan, B. A Scientific Conception of Animal Welfare that Reflects Ethical Concerns. *Anim. Welf.* **1997**, *6*, 187–205.
32. Korte, S.M.; Olivier, B.; Koolhaas, J.M. A new animal welfare concept based on allostasis. *Physiol. Behav.* **2007**, *92*, 422–428. [CrossRef]
33. Sneddon, L.U.; Wolfenden, D.C.C.; Thomson, J.S. 12—Stress Management and Welfare. In *Fish Physiology*; Schreck, C.B., Tort, L., Farrell, A.P., Brauner, C.J., Eds.; Biology of Stress in Fish; Academic Press: Cambridge, MA, USA, 2016; Volume 35, pp. 463–539.

34. Bain, M.B.; Webb, D.H.; Tangedal, M.D.; Mangum, L.N. Movements and Habitat Use by Grass Carp in a Large Mainstream Reservoir. *Trans. Am. Fish. Soc.* **1990**, *119*, 553–561. [CrossRef]

35. Kohda, M. Territoriality of male cichlid fishes in Lake Tanganyika. *Ecol. Freshw. Fish* **1995**, *4*, 180–184. [CrossRef]

36. McConnell, R.H.L. Breeding behaviour patterns and ecological differences between tilapia species and their significance for evolution within the genus tilapia (Pisces: Cichlidae). *Proceedings of the Zoological Society of London* **1959**, *132*, 1–30. [CrossRef]

37. Abecasis, D.; Erzini, K. Site fidelity and movements of gilthead sea bream (*Sparus aurata*) in a coastal lagoon (Ria Formosa, Portugal). *Estuar. Coast. Shelf Sci.* **2008**, *79*, 758–763. [CrossRef]

38. Arechavala-Lopez, P.; Uglem, I.; Fernandez-Jover, D.; Bayle-Sempere, J.T.; Sanchez-Jerez, P. Post-escape dispersion of farmed seabream (*Sparus aurata* L.) and recaptures by local fisheries in the Western Mediterranean Sea. *Fish. Res.* **2012**, *121–122*, 126–135. [CrossRef]

39. Bruton, M.N. The Habitats and Habitat Preferences of Clarias Gariepinus (pisces: Clariidae) in a Clear Coastal Lake (lake Sibaya, South Africa). *J. Limnol. Soc. S. Afr.* **1978**, *4*, 81–88.

40. Hocutt, C.H. Seasonal and diel behaviour of radio-tagged Clarias gariepinus in Lake Ngezi, Zimbabwe (Pisces: Clariidae). *J. Zool.* **1989**, *219*, 181–199. [CrossRef]

41. Wood-Gush, D.G.M.; Vestergaard, K. Exploratory behavior and the welfare of intensively kept animals. *J. Agric. Ethics* **1989**, *2*, 161–169. [CrossRef]

42. Huntingford, F.A.; Kadri, S. Defining, assessing and promoting the welfare of farmed fish. *Rev. Sci. Tech. OIE* **2014**, *33*, 233–244. [CrossRef]

43. Van de Vis, J.W.; Poelman, M.; Lambooij, E.; Bégout, M.-L.; Pilarczyk, M. Fish welfare assurance system: Initial steps to set up an effective tool to safeguard and monitor farmed fish welfare at a company level. *Fish Physiol. Biochem.* **2012**, *38*, 243–257. [CrossRef] [PubMed]

44. Williot, P.; Chebanov, M.; Nonnotte, G. Welfare in the Cultured Siberian Sturgeon, Acipenser baerii Brandt: State of the Art. In *The Siberian Sturgeon (Acipenser baerii, Brandt, 1869) Volume—Farming*; Springer: Cham, Switzerland, 2018; pp. 403–450. ISBN 978-3-319-61674-2.

45. Filiciotto, F.; Giacalone, V.M.; Fazio, F.; Buffa, G.; Piccione, G.; Maccarrone, V.; Di Stefano, V.; Mazzola, S.; Buscaino, G. Effect of acoustic environment on gilthead sea bream (*Sparus aurata*): Sea and onshore aquaculture background noise. *Aquaculture* **2013**, *414–415*, 36–45. [CrossRef]

46. Bullock, A.M. Solar Ultraviolet Radiation: A Potential Environmental Hazard in the Cultivation of Farmed Finfish. In *Recent Advances in Aquaculture: Volume 3*; Muir, J.F., Roberts, R.J., Eds.; Springer Netherlands: Dordrecht, The Netherlands, 1988; pp. 139–224. ISBN 978-94-011-9743-4.

47. Bui, S.; Oppedal, F.; Sievers, M.; Dempster, T. Behaviour in the toolbox to outsmart parasites and improve fish welfare in aquaculture. *Rev. Aquac.* **2019**, *11*, 168–186. [CrossRef]

48. El Naggar, G.O.; John, G.; Rezk, M.A.; Elwan, W.; Yehia, M. Effect of varying density and water level on the spawning response of African catfish *Clarias gariepinus*: Implications for seed production. *Aquaculture* **2006**, *261*, 904–907. [CrossRef]

49. McCormick, S.D.; Hansen, L.P.; Quinn, T.P.; Saunders, R.L. Movement, migration, and smolting of Atlantic salmon (*Salmo salar*). *Can. J. Fish. Aquat. Sci.* **1998**, *55*, 77–92. [CrossRef]

50. Huntingford, F.A.; Adams, C.; Braithwaite, V.A.; Kadri, S.; Pottinger, T.G.; Sandøe, P.; Turnbull, J.F. Current issues in fish welfare. *J. Fish Biol.* **2006**, *68*, 332–372. [CrossRef]

51. *Methods in Reproductive Aquaculture: Marine and Freshwater Species*; Marine Biology Series; Cabrita, E.; Robles, V.; Herráez, P. (Eds.) CRC Press: Boca Raton, FL, USA, 2009; ISBN 978-0-8493-8053-2.

52. Conte, F.S. Stress and the welfare of cultured fish. *Appl. Anim. Behav. Sci.* **2004**, *86*, 205–223. [CrossRef]

53. Ellis, T.; Scott, S.; Bromage, N.; North, B.; Porter, M. What is stocking density? *Trout News* **2001**, *32*, 35–37.

54. Turnbull, J.F.; North, B.P.; Ellis, T.; Adams, C.E.; Bron, J.; MacIntyre, C.M.; Huntingford, F.A. Stocking density and the welfare of farmed salmonids. *Fish Welf.* **2008**, 111–120. [CrossRef]

55. Gilmour, K.M.; DiBattista, J.D.; Thomas, J.B. Physiological Causes and Consequences of Social Status in Salmonid Fish. *Integr. Comp. Biol.* **2005**, *45*, 263–273. [CrossRef] [PubMed]

56. Sørensen, C.; Nilsson, G.E.; Summers, C.H.; Øverli, Ø. Social stress reduces forebrain cell proliferation in rainbow trout (Oncorhynchus mykiss). *Behav. Brain Res.* **2012**, *227*, 311–318. [CrossRef]

57. Cammarata, M.; Vazzana, M.; Accardi, D.; Parrinello, N. Seabream (*Sparus aurata*) long-term dominant-subordinate interplay affects phagocytosis by peritoneal cavity cells. *Brain Behav. Immun.* **2012**, *26*, 580–587. [CrossRef]

58. Sloman, K.A.; Metcalfe, N.B.; Taylor, A.C.; Gilmour, K.M. Plasma cortisol concentrations before and after social stress in rainbow trout and brown trout. *Physiol. Biochem. Zool.* **2001**, *74*, 383–389. [CrossRef]

59. Sloman, K.A.; Montpetit, C.J.; Gilmour, K.M. Modulation of catecholamine release and cortisol secretion by social interactions in the rainbow trout, Oncorhynchus mykiss. *Gen. Comp. Endocrinol.* **2002**, *127*, 136–146. [CrossRef]

60. Olsen, R.E.; Ringø, E. Dominance hierarchy formation in Arctic charr *Salvelinus alpinus* (L.): Nutrient digestibility of subordinate and dominant fish. *Aquac. Res.* **1999**, *30*, 667–671. [CrossRef]

61. Hatlen, B.; Grisdale-Helland, B.; Helland, S.J. Growth variation and fin damage in Atlantic cod (*Gadus morhua* L.) fed at graded levels of feed restriction. *Aquaculture* **2006**, *261*, 1212–1221. [CrossRef]

62. Faisal, M.; Chiappelli, F.; Ahmed, I.I.; Cooper, E.L.; Weiner, H. Social confrontation "Stress" in aggressive fish is associated with an endogenous opioid-mediated suppression of proliferative response to mitogens and nonspecific cytotoxicity. *Brain Behav. Immun.* **1989**, *3*, 223–233. [CrossRef]

63. Khater, E.G. *Simulation Model for Design and Management of Water Recirculating Systems in Aquaculture*; Agricultural Engineering Department, Faculty of Agriculture, Moshtohor, Benha University: Moshtohor, Egypt, 2012.

64. Malone, R. *Recirculating Aquaculture Tank Production Systems*; USDA, Southern Regional Aquaculture Center: Stoneville, MS, USA, 2013; p. 12.

65. Powell, A.; Treasurer, J.W.; Pooley, C.L.; Keay, A.J.; Lloyd, R.; Imsland, A.K.; de Leaniz, C.G. Use of lumpfish for sea-lice control in salmon farming: Challenges and opportunities. *Rev. Aquac.* **2018**, *10*, 683–702. [CrossRef]

66. Daborn, G.R.; Gregory, R.S. Occurrence, distribution, and feeding habits of juvenile lumpfish, *Cyclopterus lumpus* L. in the Bay of Fundy. *Can. J. Zool.* **1983**, *61*, 797–801. [CrossRef]

67. Lines, J.A.; Spence, J. Safeguarding the welfare of farmed fish at harvest. *Fish Physiol. Biochem.* **2012**, *38*, 153–162. [CrossRef] [PubMed]

68. Boglione, C.; Gisbert, E.; Gavaia, P.; Witten, P.; Moren, M.; Fontagné, S.; Koumoundouros, G. Skeletal anomalies in reared European fish larvae and juveniles. Part 2: Main typologies, occurrences and causative factors. *Rev. Aquac.* **2013**, *5*, S121–S167. [CrossRef]

69. Boglione, C.; Gavaia, P.; Koumoundouros, G.; Gisbert, E.; Moren, M.; Fontagné, S.; Witten, P.E. Skeletal anomalies in reared European fish larvae and juveniles. Part 1: Normal and anomalous skeletogenic processes. *Rev. Aquac.* **2013**, *5*, S99–S120. [CrossRef]

70. Soares, F.; Fernández, I.; Costas, B.; Gavaia, P.J. Non-infectious disorders of warmwater fish. In *Diseases and Disorders of Finfish in Cage Culture*; CAB International: Wallingford, UK, 2014.

71. European Food Safety Authority (EFSA). Species-specific welfare aspects of the main systems of stunning and killing of farmed Seabass and Seabream: Species-specific welfare aspects of the main systems of stunning and killing of farmed Seabass and Seabream. *EFSA J.* **2009**, *1010*, 1–52.

72. Digre, H.; Erikson, U.; Misimi, E.; Lambooij, B.; Van De Vis, H. Electrical stunning of farmed Atlantic cod *Gadus morhua* L.: A comparison of an industrial and experimental method. *Aquac. Res.* **2010**, *41*, 1190–1202.

73. European Food Safety Authority (EFSA). Species-specific welfare aspects of the main systems of stunning and killing of farmed Atlantic Salmon. *EFSA J.* **2009**, *1012*, 1–77.

74. European Food Safety Authority (EFSA). Species-specific welfare aspects of the main systems of stunning and killing of farmed fish: Rainbow Trout. *EFSA J.* **2009**, *7*. [CrossRef]

75. European Food Safety Authority (EFSA). Species-specific welfare aspects of the main systems of stunning and killing of farmed Carp. *EFSA J.* **2009**, *1013*, 1–37.

76. European Food Safety Authority (EFSA). Species-specific welfare aspects of the main systems of stunning and killing of farmed tuna. *EFSA J.* **2009**, *1072*, 1–53.

77. European Food Safety Authority (EFSA). Scientific opinion of the panel on animal health and welfare on a request from the European commission on animal welfare aspects of husbandry systems for farmed European seabass and gilthead seabream. *EFSA J.* **2008**, *844*, 1–21.

78. Lambooij, E.; Gerritzen, M.A.; Reimert, H.; Burggraaf, D.; van de Vis, J.W. A humane protocol for electro-stunning and killing of Nile tilapia in fresh water. *Aquaculture* **2008**, *275*, 88–95. [CrossRef]

79. Robb, D.H.F.; Kestin, S.C. Methods Used to Kill Fish: Field Observations and Literature Reviewed. *Anim. Welf.* **2002**, *11*, 269–282.

80. Tidwell, J.H.; Allan, G.L. Fish as food: aquaculture's contribution. Ecological and economic impacts and contributions of fish farming and capture fisheries. *EMBO Rep.* **2001**, *2*, 958–963. [CrossRef]

81. Turchini, G.M.; Torstensen, B.E.; Ng, W.-K. Fish oil replacement in finfish nutrition. *Rev. Aquac.* **2009**, *1*, 10–57. [CrossRef]

82. Jensen, U. Probabilistic Risk Analysis: Foundations and Methods. *J. Am. Stat. Assoc.* **2002**, *97*, 925–926. [CrossRef]

83. Teletchea, F.; Fontaine, P. Levels of domestication in fish: Implications for the sustainable future of aquaculture. *Fish Fish.* **2012**, *15*, 181–195. [CrossRef]

84. Teletchea, F. Domestication of Marine Fish Species: Update and Perspectives. *J. Mar. Sci. Eng.* **2015**, *3*, 1227–1243. [CrossRef]

85. Huntingford, F.A.; Kadri, S. Welfare and fish. *Fish Welf.* **2008**, *1*, 19–32.

86. Poli, B.M.; Parisi, G.; Scappini, F.; Zampacavallo, G. Fish welfare and quality as affected by pre-slaughter and slaughter management. *Aquac. Int.* **2005**, *13*, 29–49. [CrossRef]

87. Morzel, M.; Sohier, D.; Van de Vis, H. Evaluation of slaughtering methods for turbot with respect to animal welfare and flesh quality. *J. Sci. Food Agric.* **2003**, *83*, 19–28. [CrossRef]

88. Napolitano, F.; Girolami, A.; Braghieri, A. Consumer liking and willingness to pay for high welfare animal-based products. *Trends Food Sci. Technol.* **2010**, *21*, 537–543. [CrossRef]

89. Honkanen, P.; Olesen, I.; Mejdell, C.; Nielsen, H.M.; Grimsrud, K.; Gamborg, C.; Navrud, S.; Ellingsen, K.; Sandøe, P. Who cares about fish welfare? A Norwegian study. *Br. Food J.* **2015**, *117*, 257–273.

90. Ottar Olsen, S.; Honkanen, P. Environmental and animal welfare issues in food choice: The case of farmed fish. *Br. Food J.* **2009**, *111*, 293–309.

91. Adams, C.E.; Turnbull, J.F.; Bell, A.; Bron, J.E.; Huntingford, F.A. Multiple determinants of welfare in farmed fish: Stocking density, disturbance, and aggression in Atlantic salmon (*Salmo salar*). *Can. J. Fish. Aquat. Sci.* **2007**, *64*, 336–344. [CrossRef]

92. Foss, A.; Grimsbø, E.; Vikingstad, E.; Nortvedt, R.; Slinde, E.; Roth, B. Live chilling of Atlantic salmon: Physiological response to handling and temperature decrease on welfare. *Fish Physiol. Biochem.* **2012**, *38*, 565–571. [CrossRef] [PubMed]

93. Cañon Jones, H.A.; Noble, C.; Damsgård, B.; Pearce, G.P. Investigating the influence of predictable and unpredictable feed delivery schedules upon the behaviour and welfare of Atlantic salmon parr (Salmo salar) using social network analysis and fin damage. *Appl. Anim. Behav. Sci.* **2012**, *138*, 132–140. [CrossRef]

94. Noble, C.; Berrill, I.; Waller, B.; Kankainen, M.; Setala, J.; Honkanen, P.; Mejdell, C.M.; Turnbull, J.; Damsgard, B.; Schneider, O.; et al. A multi-disciplinary framework for bio-economic modeling in aquaculture: A welfare case study. *Aquac. Econ. Manag.* **2012**, *16*, 297–314. [CrossRef]

MDPI

St. Alban-Anlage 66

4052 Basel

Switzerland

Tel. +41 61 683 77 34

Fax +41 61 302 89 18

www.mdpi.com

*Fishes* Editorial Office

E-mail: fishes@mdpi.com

www.mdpi.com/journal/fishes

www.ingramcontent.com/pod-product-compliance
Lightning Source LLC
Chambersburg PA
CBHW051911210326
41597CB00033B/6110